Indoor Air Research Series

Max Eisenberg
Series Editor

ASTHMA: CAUSES and MECHANISMS of an EPIDEMIC INFLAMMATORY DISEASE

Edited by
Thomas Platts-Mills

Center for Indoor Air Research

CIAR

Lewis Publishers
Boca Raton London New York Washington, D.C.

Library of Congress Cataloging-in-Publication Data

Asthma : causes and mechanisms of an epidemic inflammatory disease / edited by Thomas Platts-Mills.
p. cm. -- (Indoor air research series)
ISBN 1-56670-373-5 (alk. paper)
1. Asthma--Etiology. 2. Asthma--Pathophysiology. I. Platts-Mills, Thomas. II. Series.
[DNLM: 1. Asthma. 2. Air Pollution, Indoor--adverse effects. WF 553 A85305 1999]
RC591.A755 1999
616.2′38—dc21
DNLM/DLC
for Library of Congress 99-17632
CIP

Lewis Publishers is an imprint of CRC Press LLC

International Standard Book Number 1-56670-373-5
Library of Congress Card Number 99-17632
Printed in the United States of America 1 2 3 4 5 6 7 8 9 0
Printed on acid-free paper

Series Preface

The field of indoor air science is of growing interest and concern given that modern society spends the better part of each day indoors. Since the indoor air environment is a major, continual exposure medium for occupants, it is important to study what is present and if and how it affects the health and comfort of occupants.

Volumes in this Indoor Air Research Series are intended to provide state-of-the-art information on many areas germane to indoor air science including Chemical and Biological Sources, Exposure Assessment, Dosimetry, Engineering Controls, and Perception of Indoor Air Quality. In each volume, authors known for their expertise on the topic will present comprehensive and critical accounts of our current understanding in the area.

It is hoped that the series will advance knowledge and broaden interest among the scientific community at large in the indoor air science field.

Max Eisenberg, Ph.D.
Series Editor

Preface

The beginning of the 20th Century was characterized by major epidemics of tuberculosis, syphilis, and influenza which killed or maimed millions of people. By 1940 the major infectious diseases of the Western world had been conquered. However, as we approach the end of the century, a different group of diseases has increased. For some of the diseases that have increased, e.g. coronary artery disease, carcinoma of the colon and Type II diabetes, there appear to be good explanations in terms of changes in lifestyle or diet. By contrast, the reasons for the increase in asthma are not clear. Several recent conferences have failed to reach any consensus about the causes of the increase in asthma. Increasingly we do understand the mechanisms by which the lungs can become inflamed including the contribution of lymphocytes, antibodies, mast cells, eosinophils, the epithelial layer and blood vessels within the bronchial walls. The challenge is to understand the causes of inflammation and thus to understand the reasons for the increase. However, asthma is not just a simple inflammatory disease. There are still major questions about the relationship between inflammation, bronchial hyperreactivity and symptoms. In order to understand the increase in asthma it is essential to step back and consider the ways in which lifestyle changes could influence immunity, immunopathology and physiology of the lungs.

The function of the lung involves exposure to approximately one cubic meter of air per hour. This air contains noxious gases; pathogens (bacteria, viruses and fungal spores); and a wide range of non-viable particles, some of which carry foreign proteins. Although only the foreign proteins give rise to IgE antibodies and T cell responses, the contents of the air can contribute to inflammation and asthma in many different ways. For example, viruses can trigger acute attacks and positively interact with allergens; similarly, ozone can increase the response to allergens and tobacco smoke can produce both acute triggering effects and more prolonged effects on the lungs. The problem is to understand how these factors come together to produce disease and which effects are involved in the increasing prevalence and severity of the disease. At present it is very difficult to identify a smoking gun in terms of a simple agent that can cause or trigger asthma and has increased consistently in so many different countries. The scale of the current confusion can be gauged from the fact that in England some scientists are blaming western cleanliness for the increase in asthma, while in the United States the worst epidemic involves city children living in apartments infested with rats and cockroaches. However, in almost all countries where asthma has increased, the strongest correlation is with immediate hypersensitivity to proteins found inside houses, i.e. those derived from dust mites, domestic animals and the German cockroach. Indeed there is very strong evidence that dust mite and other indoor allergens are an important *cause* of asthma. Thus increasing evidence has focused on reasons why the response of the lungs to indoor allergens should have increased, or put another way, to look for changes in lifestyle or diet that could have lead to the loss of an effect that protected the lungs against wheezing.

This book brings together the views of a group of scientists addressing asthma research from many different directions. The object was to provide a broad view of the research that is being done but also to put it together so that each aspect can be considered in the light of the ongoing epidemic of inflammatory disease that is proving very resistant both to treatment and to scientific analysis.

Thomas A.E. Platts-Mills, M.D., Ph.D.
Editor

Acknowledgements

We would like to thank the staff of the Center for Indoor Air Research for their support and for organizing the conference. In particular we wish to thank Alice Zeiger who has been with the project from its inception and Max Eisenberg for his unfailing enthusiasm. The contributors are (or should be) grateful to Nancy Malone for her patience and skill in preparing the manuscripts.

The Editor

Professor Thomas Platts-Mills was born in England, received a BA in Animal Physiology from Balliol College, Oxford; Medical Qualification from St. Thomas Hospital in London; and a Ph.D. in Immunology from London University. Following internal medicine training in England, he held a three-year post-doctoral fellowship from the National Institutes of Health at Johns Hopkins Medical School, working with Dr. Kimishige Ishizaka, Dr. Philip Norman and Dr. Lawrence Lichtenstein. From 1974-1984 he served on the staff of the Medical Research Council in the UK and became an Honorary Consultant Physician to Northwick Park Hospital. It was during this time that his work on dust mites and their role in allergic disease was initiated. This led to the definition of the mite allergen, Der p 1, identification of mite fecal particles as the major form in which mite allergens become airborne, and evidence that nonspecific bronchial reactivity can be reversed by allergen avoidance.

Dr. Platts-Mills' research on asthma at the University of Virginia has been supported by grants from the National Institutes of Health for 15 years. In 1993 he received a merit award from the NIH for work on the role of indoor allergens in asthma, and in the same year he was named director of the University of Virginia Asthma & Allergic Diseases Center. The research of his group has defined the role of dust mites, domestic animals and the German cockroach in asthma. This work has provided a basis for understanding the reasons for the increase in asthma and a rationale for allergen specific treatment. His funding has resulted in over 200 scholarly publications. He has served on the editorial boards of *The Journal of Allergy and Clinical Immunology, Clinical and Experimental Allergy, Clinical and Experimental Immunology, American Journal of Respiratory and Critical Care Medicine*, and the *American Journal of Medicine*. In 1995 he was elected to the Board of Directors of the American Academy of Asthma, Allergy and Immunology.

Contributors

Alessandro de Alarcon, BA
Department of Pediatrics
University of Virginia
Charlottesville, Virginia

Kathleen Barnes, PhD
Asthma & Allergy Center
Johns Hopkins University
Baltimore, Maryland

Brian Boehlecke, MD MSPH
Center for Environmental
Medicine & Lung Biology
University of North Carolina
Chapel Hill, North Carolina

Melody C. Carter, MD
Department of Family Medicine
Emory University
Atlanta, Georgia

Gang Chen, MD
Division of Allergy & Immunol.
University of Kentucky
Lexington, Kentucky

Allison D. Fryer, PhD
Asthma & Allergy Center
Johns Hopkins University
Baltimore, Maryland

Peter W. Heymann, MD
Department of Pediatrics
University of Virginia
Charlottesville, Virginia

David B. Jacoby, MD
Department of Pulmonary Drug
Discoveries
Bristol Meyers Squibb
Princeton, New Jersey

Bann C. Kang, MD
Division of Allergy & Immunol.
University of Kentucky
Lexington, Kentucky

Jin Tae Kim, MD
Division of Allergy & Immunol.
University of Kentucky
Lexington, Kentucky

Samuel B. Lehrer, PhD
Section of Clinical Immunology
Tulane University Medical Center
New Orleans, Louisiana

Mark C. Liu, MD
Asthma & Allergy Center
Johns Hopkins University
Baltimore, Maryland

Manuel Lopez, MD
Section of Clinical Immunology
Tulane University Medical Center
New Orleans, Louisiana

Beth Miller, MD
Division of Allergy & Immunol.
University of Kentucky
Lexington, Kentucky

David B. Peden, MD
Center for Environmental
Medicine & Lung Biology
University of North Carolina
Chapel Hill, North Carolina

Matthew Perzanowski, BS
Asthma and Allergic Diseases
Center
University of Virginia
Charlottesville, Virginia

Thomas A.E. Platts-Mills, MD PhD
Department of Medicine
University of Virginia
Charlottesville, Virginia

Robyn Proffitt, BS
Division of Allergy & Immunol.
University of Kentucky
Lexington, Kentucky

David Proud, PhD
Asthma & Allergy Center
Johns Hopkins University
Baltimore, Maryland

Gary P. Rakes, MD
Department of Pediatrics
University of Virginia
Charlottesville, Virginia

Roy J. Rando, ScD
Section of Clinical Immunology
Tulane University Medical Center
New Orleans, Louisiana

Lanny J. Rosenwasser, MD
Division of Allergy &
Clinical Immunology
National Jewish Center
Denver, Colorado

Scherer P. Sanders, PhD
Asthma & Allergy Center
Johns Hopkins University
Baltimore, Maryland

Alisa M. Smith, PhD
Asthma & Allergic Diseases
Center, University of Virginia
Charlottesville, Virginia

Marsha A. Wills-Karp, PhD
Department of Environmental
Health Sciences
Johns Hopkins University
Baltimore, Maryland

Judith A. Woodfolk, MB ChB
Asthma & Allergic Diseases
Center
University of Virginia
Charlottesville, Virginia

Juan C. Zambrano, MD
Department of Pediatrics
University of Virginia
Charlottesville, Virginia

Daohong Zhou, MD
Division of Allergy & Immunol.
University of Kentucky
Lexington, Kentucky

Table of Contents

Chapter 1

THE RISING PREVALENCE AND SEVERITY OF ASTHMA IN WESTERN SOCIETY: ARE THE CAUSES OF ASTHMA THE CAUSES OF THE INCREASE?

T.A.E. Platts-Mills, MD, PhD; M. Perzanowski, BS; M.C. Carter, MD; J.A. Woodfolk, MB, ChB

HISTORICAL INTRODUCTION

Alterations in the caliber of the airways can give rise to wheezing, coughing, tightness in the chest or simply shortage of breath. While there are many causes for closing of the airways, the most common association is with diffuse bronchial hyper-reactivity and the closely related eosinophil rich inflammation of the bronchial walls. The presence of eosinophils in the sputum of asthmatics was recognized before 1900. In addition, the presence of pieces of bronchial epithelium (Curshman's spirals) left little doubt that the lining of the lungs was damaged (Gleich & Adolphson, 1986). Postmortem examination of patients who died of asthma confirmed that there was extensive damage to the epithelium, infiltration with eosinophils and deposition of eosinophil derived proteins (Hogg,1992). Recent advances in bronchoscopic biopsy technique have made it possible to show that inflammation is also present in mild cases of asthma (Beasley et al, 1989). This has led to [or encouraged] major developments in three areas: 1) investigation of the mechanisms of inflammation, 2)understanding the causes of inflammation and 3) a semi-religious pressure to treat asthma with anti-inflammatory drugs. Inherent in these developments is the assumption that increased prevalence and severity of asthma is related to increased inflammation of the lungs. However, there is no conclusive evidence that the severity of asthma is a direct function of the severity of inflammation or that there has been an increase in the number of individuals with inflammation in their lungs.

In the last century asthma was considered to be a rare disease and Osler stated that "asthma never kills." Furthermore, in the 1960's asthma was considered to be more common among children, of above average intelligence, who were generally thin and who came predominantly from middle class families. Since 1960 everything has changed. First there was an epidemic of asthma deaths in the UK and New Zealand during 1966-67, which was temporally related to the introduction of a high strength isoprenaline inhaler (Jackson et al, 1988). In 1970, Morrison Smith reported that the prevalence of asthma among school children in Birmingham, England had increased over the previous 10 years. His studies also provided clear evidence that skin sensitization to dust mites was strongly

1-56670-373-5/99/$0.00+$.50

associated with asthma (Smith et al, 1969). The association between asthma and house dust sensitization had been reported in 1921 by Kern (Kern, 1921) and the importance of dust mites of the genus Dermatophagoides, as a source of house dust allergens, was established by Voorhorst and his colleagues in Holland (Spieksma & Voorhorst, 1969). That group also reported that there were major differences in the number of mites found in house dust from different climatic areas (Voorhorst et al, 1969). In 1970 Kerrebijn presented the results of repeated bronchial hyper-reactivity testing on mite allergic children during a year spent at a sanatorium in Davos, Switzerland (altitude ~6,000 ft,) (Kerrebijn, 1970). As a group, the children improved clinically and experienced highly significant decreases in BHR to histamine. Indeed, half the children returned to the normal range for bronchial reactivity. In that same year Altounyan reported the evidence that seasonal exposure to pollen could increase non-specific BHR and that this would generally return to normal within three months after the pollen season (Altounyan, 1970). Thus, by 1971 it could be said that there was evidence for an increase in asthma among mite allergic children and also evidence that full avoidance of exposure (i.e. in a sanatorium) could decrease symptoms and BHR. Furthermore, it was already clear that most of these children had very nonspecific perennial symptoms with no clear relationship to exposure, in keeping with the view that chronic exposure to mite allergens was an important cause of bronchial inflammation and BHR, rather than being a trigger for asthma.

During the 1970's and early 1980's evidence for increased prevalence of asthma continued to accumulate. This evidence came from countries such as New Zealand, Australia, Japan, Holland, and the UK, where dust mite sensitization was very common among children with asthma. In addition, there was a striking correlation with changes in houses that had occurred between 1955 and 1975 which could have increased the growth of dust mites. The main changes were increased temperature, decreased ventilation, the widespread adoption of fitted carpeting and increased upholstered furniture. It was also clear that children were spending more time indoors which could have further increased exposure to house dust related allergens (Platts-Mills et al, 1984). In 1978 the first dust mite allergen was purified, which made it possible to quantitate mite allergen in samples of house dust or airborne (Chapman & Platts-Mills, 1980). The airborne studies established that allergen was only airborne during disturbance and that it became airborne predominantly in the form of mite fecal particles (Tovey et al, 1981a; Tovey et al, 1981b). These particles are 10 - 30 microns in diameter (i.e. similar in size to pollen grains) and fall rapidly after disturbance. In 1981 we reported that young patients admitted to special "allergen free" hospital rooms experienced decreased symptoms, decreased medication requirement and decreased BHR to histamine (Platts-Mills et al, 1982). Taken together, the evidence available at that time led to three conclusions:

1. Chronic exposure to dust mite allergens was an important cause of sensitization and asthma
2. Effective decrease in dust mite allergens can decrease both the symptoms of asthma and nonspecific BHR
3. Increased exposure to dust mites, due to changes in housing and life style, was a probable cause of the increase in prevalence and severity of asthma

Over the last 15 years there have been major changes in our understanding of the mechanisms by which allergens and other inhaled chemicals or proteins can cause inflammation of the bronchi. It has become well accepted that allergens can produce immune responses, including T cells of the T_H2 type, IgE antibodies and IgG4 antibodies, as well as recruitment of eosinophils to the lungs (Wierenga et al, 1990; Aalberse et al, 1983). The argument that dust mites are an important cause of asthma has become progressively stronger (Sporik et al, 1992) (Table I). On the other hand, there has been a major change in the evidence about the epidemic such that it is no longer possible to argue that increases in dust mite exposure are the primary cause of the increase in asthma.

Table I

Hill's criteria for causality and the evidence supporting a causal link between mite allergen exposure and the development of asthma

1) The strength of the association is large: (i) population; (ii) case control; and (iii) prospective studies.
2) Repeated observations in different populations have provided consistent findings: UK, Europe, Australia, USA, and others.
3) A cause leads to a specific effect, i.e. asthma and not other lung diseases.
4) There is a dose response gradient between exposure and sensitization.
5) Experimental evidence favoring causality from: (a) avoidance studies; (b) challenge studies.
6) There are analogous explanations and the mechanism is biologically plausible.

*Hill AB, Proc Roy Soc Med 1965;58:295. Sporik RB, *et al*, Clin Exp Allergy 1992;22:897.

A. THE RELEVANCE OF ALLERGEN SENSITIZATION TO ASTHMA

When allergen extracts are injected into the skin of a non-allergic person there is no significant immediate or delayed response in the skin. Similarly, the immediate and late airway response to bronchial challenge is restricted to allergic individuals. In epidemiological studies it is sensitization to allergens that has been consistently associated with asthma. Thus from both the direct provocation experiments and epidemiology it appears that the role of allergens in causing inflammation in the lungs is restricted to individuals who have made an immune response (Sporik et al, 1992; Shaver et al, 1997;

Liu et al, 1991; Calhoun et al, 1994). For dust mite allergens there are studies from several different countries showing that sensitization is directly related to the concentration of allergens in the patient's house (Platts-Mills et al, 1992). Sensitization to inhalant allergens is strongly influenced by genetics. The proposed "threshold" for dust mite allergen exposure [i.e. >2 ug Group I mite allergen/gram of dust] is the threshold for genetically predisposed individuals (Platts-Mills et al, 1992; Platts-Mills et al, 1997). From the results of Kuehr *et al* it appears that sensitization of genetically "non-allergic" individuals requires concentrations as high as 60 ug Group I mite allergen/gram of dust. [The results could also be interpreted as showing that there is a gradient of genetically predisposed individuals, so that with higher levels of exposure, a progressively larger proportion of the population would be judged to be atopic (Kuehr et al, 1994)]. Sensitization to other indoor allergens is also dependent on exposure, but the evidence for a dose response relationship is much less clear. In a community where 50% or more of the houses contain animals, all children are exposed to significant cat (or dog) allergens, and the presence of an animal in the child's home is not a good predictor of sensitization. The reason for this appears to be that animal dander (and allergen) is carried on clothing and becomes widespread in the community. Significant levels of cat (and dog) allergen are found both in schools and in houses where there is no animal (Gelber et al, 1993; Ingram et al, 1995; Perzanowski et al, 1999, in press; Ronmark et al, 1998). By contrast, in communities where less than 10% of the families have indoor animals [i.e. the inner city population of Atlanta or the African American community in Charlottesville] the concentration of cat allergen in house dust is generally <1 ug Fel d 1/g dust and sensitization is uncommon (Call et al, 1992; Rosenstreich et al, 1997). Sensitization to cockroach allergens is also restricted to those communities where many of the houses are infested. Immediate sensitivity to proteins derived from *Blattella germanica* is common among asthmatics in most large American cities (Call et al, 1992; Rosenstreich et al, 1997; Kang et al, 1979; Bernton et al, 1972) but very rare among patients or controls living in suburban areas. Overall, the conclusion is that the pattern of sensitization among asthmatics can be predicted from the mean concentration of allergen in the houses where they live (Squillace et al, 1997). Only for dust mite allergen has a direct relationship been demonstrated between sensitization and the concentration in the individual patient's house.

As we have already stated, the demonstrated effects of allergen exposure are restricted to allergic individuals. However, in all case-control or population based studies there are many allergic individuals who do not report symptoms of asthma and who do not have BHR (Squillace et al, 1997; Sporik et al, 1990; Peat et al, 1996). Thus sensitization to indoor allergens is a risk factor for asthma in an epidemiological sense. The question remains as to why some allergic individuals have symptoms of asthma, while others have rhinitis and yet others report no symptoms.

Initially we assumed that the concentration of allergen in the allergic patient's house would be the primary determinant of symptoms (Sporik et al, 1992; Platts-Mills et al, 1992; Gelber et al, 1993). In some studies a significant relationship between exposure and symptoms has been observed. However, in many others this does not appear to be true. The implication is that once an individual has developed immediate hypersensitivity, the quantity of allergen necessary to induce symptoms is very variable. (See Detailed discussion in Platts-Mills et al, 1995). Understanding the factors that influence the severity of symptoms among allergic individuals becomes an important issue, both in relation to treatment and in understanding the reasons for the increase in asthma.

B. WHAT CONTROLS THE PRESENCE AND SEVERITY OF SYMPTOMS AMONG ALLERGIC INDIVIDUALS

The scale of this question is enormous. In many studies up to 30% of the population are allergic as judged by skin tests while only half the allergic population have clinically important symptoms, and ~ one quarter (i.e. ~5% of the total population) have asthma as defined as symptomatic BHR. Thus even though sensitization to one or more of the indoor allergens is a major risk factor for asthma, only a minority of the allergic individuals have the disease. An arbitrary classification of the factors influencing symptoms recognizes sensitizers (or inducers), enhancers and triggers (Fig.1). Sensitizers are foreign proteins, or smaller molecular weight chemicals, that are known to induce an immune response including TH2 cells and IgE antibodies. All the inhalant allergens, and some non-inhaled allergens [e.g. foods, Aspergillus and dermatophytes] are in this category. Repeated exposure to a sensitizer can give rise to local eosinophil responses in the skin or lungs. Enhancers are agents present in the environment that can increase or even cause inflammation in the lungs but have not been shown to induce TH2 responses or IgE antibodies. The best established examples are rhinoviruses, endotoxin, ozone and diesel particulates (Platts-Mills et al, 1997). For the first three of these there are clear experiments showing a positive interaction with allergen exposure. Thus rhinovirus infection can upregulate the bronchial inflammatory response to allergen challenge (Calhoun et al, 1994; Lemanske et al, 1989). Endotoxin exposure has been reported to be an important determinant of symptoms among mite allergic asthmatic patients (Michel et al, 1996). The inflammatory response of the nose and lungs to allergen can be upregulated by preceding exposure to ozone (Molfino et al, 1991; Peden et al, 1995). Whether beta-2-agonists should be regarded as an enhancer is open to question. The evidence about regular beta-2-agonist use increasing BHR is mixed (Sears et al, 1990; Tattersfield, 1994; Cockcroft, 1995). However, Cockcroft and his colleagues have demonstrated that regular treatment with albuterol can increase the inflammatory response to subsequent allergen challenge (Cockcroft, 1995). The results for diesel particulates are again different.

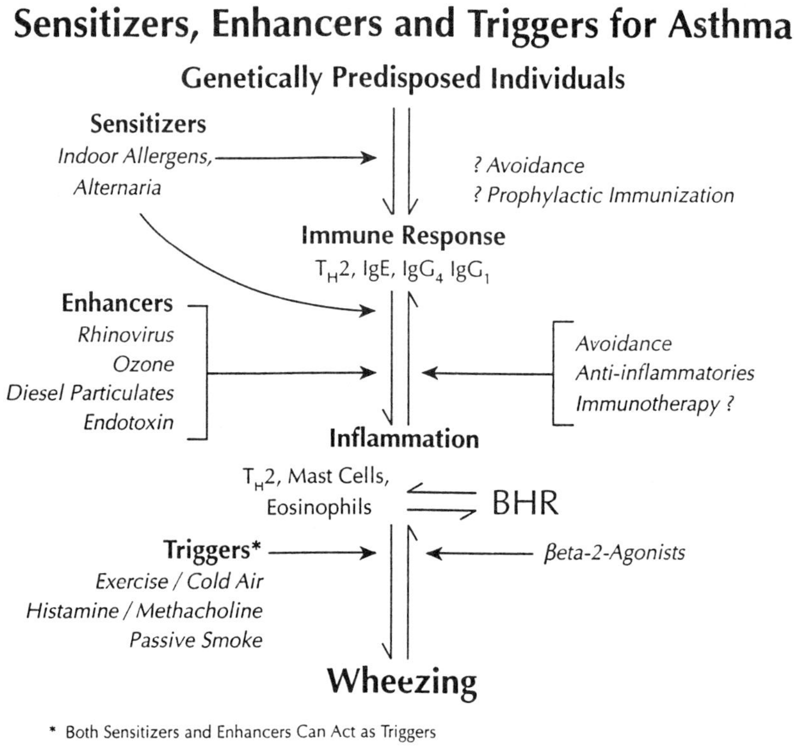

Figure 1. Sensitizers are those foreign proteins that can give rise to an immune response of the T_H2 type and can contribute to eosinophil rich inflammation. Enhancers are a special group of agents that can upregulate the inflammatory response to sensitizers but are not thought to induce an immune response in their own right. Triggers can give rise to acute episodes of wheezing but do not have any effect on inflammation.

In this case the particulates can induce an inflammatory response but they have not been shown to produce a T_H2 response and to date a positive interaction with allergen exposure has not been reported (Diaz-Sanchez et al, 1994).

Traditional triggers are those agents that can induce increased airway resistance within 20 minutes without a subsequent inflammatory response. Obvious examples are histamine, methacholine, distilled water and hypertonic saline. In addition, there is the stimulus created by aerobic exercise in dry air, or surrogates for exercise, such as cold air challenge. Passive smoke exposure is often included as a trigger. However, recent evidence that the pulmonary response to smoke may take 40 minutes suggests that there may also be an inflammatory or cytokine response to this exposure (Rando et al, 1992; Menon et al, 1992). The important feature of the true triggers, i.e. histamine, methacholine or exercise, is that there is no eosinophil response or late reaction. Thus it is possible to carry out repeated histamine or exercise challenges to monitor the effects of treatment. Although it has been difficult to demonstrate release of mediators during exercise, recent data establishing a protective effect of leukotriene antagonists strongly suggests that leukotrienes, and perhaps some other mediators, are released. By contrast, allergen challenges can trigger wheezing within 20 minutes but will also give rise to late reactions and an increase in BHR which can be prolonged. In theory an increase in any of these factors could lead to an increase in the severity of asthma. However, it is generally assumed that neither triggers nor enhancers can give rise to asthma, and that they will only act on an already predisposed population. Using the model in Fig. 1, the question can be posed: "Has the increase in asthma involved: (i) more individuals becoming allergic; (ii) more of the allergic individuals becoming inflamed; or (iii) an increase of symptoms among individuals who are already inflamed?"

C. 1980 TO THE PRESENT: THE TARGET CHANGES

Since 1980 there have been several major developments that have changed our interpretation of the relationship between allergen exposure and the increase in asthma.

1. Studies based on populations, hospital admissions, or prescriptions for asthma treatment, have found a progressive increase in cases since 1980. Indeed recent data from England and Australia suggests that the increase is continuing into the 1990's. The changes in houses that could have increased allergen exposure were basically complete by 1975. As has been pointed out, mite allergen numbers in English houses were already high in 1975 (Seaton et al, 1994). Indeed we have published data documenting that mite allergen levels in a cohort of English houses did not increase between 1979 and 1990 (Sporik et al, 1990). Thus asthma prevalence has increased steadily for almost 20 years after the main changes in houses in the UK, Australia and New Zealand that would have increased mite allergen exposure were completed. The main decreases in ventilation rates of houses also occurred prior to 1975 (Fig. 2). Thus the time course of increases in asthma prevalence no longer parallels changes in houses.

2. Increases in asthma and/or high prevalence of asthma have been documented in geographic areas where dust mites are not relevant. The most striking example is in Scandinavia where an increase in asthma prevalence has been documented among military conscripts both in Finland and Sweden (Haahtela et al, 1990; Aberg, 1989). Furthermore the increase has been at least as great in Northern Sweden as in the south (Lundback, 1993). Thus the increase has occurred in areas where the humidity is too low for mite growth, and the most important indoor allergens are those derived from domestic animals (Ronmark et al, 1998). Similarly among school children in the mountain states of the USA, the prevalence of asthma may be as high as 6% although there is no significant mite (or cockroach) allergen in the children's houses (Ingram et al, 1995; Sporik et al, 1995).

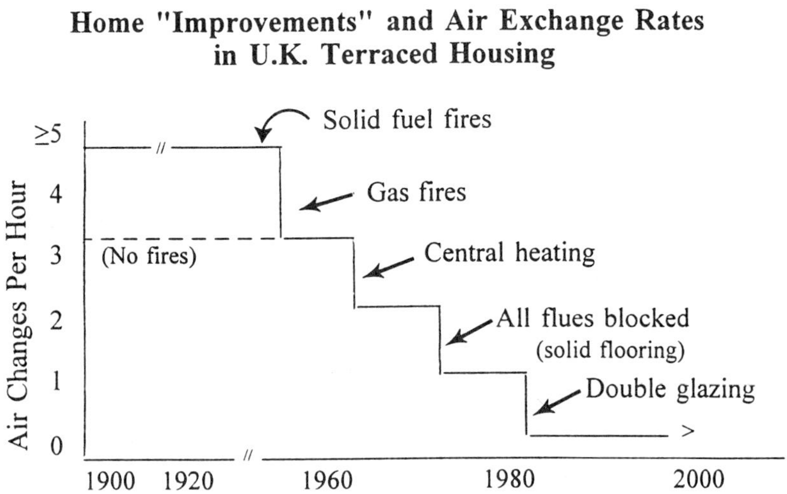

Figure 2. Air exchange rates in English houses decreased dramatically between 1950 and 1975; analysis based on data provided by Ian Ashmore and Ian Cousel [Environmental Health and Building Control, Swindon England, 1996]. However there has been little change since 1980 and changes of this magnitude did not occur in the USA.

Again, the dominant sources of allergens are cats and dogs. Another example is the importance of cockroach allergen among low income patients in American cities. It had been known for years that cockroach sensitization was common among asthmatics in Boston, Chicago, Kansas City and New York (Call et al, 1992; Rosenstreich et al, 1997; Kang et al, 1979; Bernton et al, 1972; Squillace et al, 1997). When it became possible to measure allergens from *B. germanica* it was clear that this sensitization was restricted to communities where cockroach allergen was common in the houses (Gelber et al, 1993; Call et al, 1992). In houses in Atlanta cockroach allergen is usually present together with mite allergen and sensitization to either allergen is common among asthmatic children (Call et al, 1992; Carter et al, 1997). This is in keeping with year round humidity in the state of Georgia. By contrast, in the big cities of the Northeast most poor people live in apartments which are overheated and are too dry for mite growth during 4-6 months of the year. In these cities cockroaches, and not dust mites, are the dominant source of allergens associated with asthma (Rosenstreich et al, 1997). Thus we have clear examples of increases in asthma associated with sensitization to cat and dog allergens and also to cockroach allergens.

3. Asthma has not only increased in prevalence but has also become a more difficult disease to treat. Despite the introduction of a variety of new medicines [e.g. delayed release theophylline, cromolyn sodium and topical steroids] there has been a progressive increase in hospital admissions for asthma (Mitchell & Newacheck, 1986). In controlled trials each of these medicines produces significant improvement in the disease. Similarly, although immunotherapy for asthma can produce significant improvement as judged by mean peak flow values, the results in perennial asthma are disappointing. Equally, allergen avoidance, which is the logical first line of anti-inflammatory treatment, can produce good results but does not cure the disease. Indeed none of the presently available treatments has produced major shifts in severity of this disease in a community or population.

4. Increases in asthma and allergic disease have not been restricted to the age group in which inhalant allergens are the dominant cause. From hospital admission data, there has been an increase in asthma among children under 4 years old and also among adults (Gergen & Weiss, 1990). These increases are unlikely to be attributable to increased allergen exposure. However, it is important to remember that there is very little population based data related to those age groups. There have also been reports of increases in inflammatory diseases of the upper respiratory tract. In particular, seasonal allergic rhinitis and chronic sinusitis. The evidence about increasing prevalence of hay fever is confusing. Hay fever first became epidemic in the later part of the 19th century. In 1916 a large study on the genetics of hay fever was published in the United States (Cooke & Van Der Veer, 1916). In the 1960's studies on the East Coast and in the Midwest already reported that as many as 20% of college students had symptomatic hay fever (Hagy & Settipane, 1971; Broder et al, 1974), and it is not clear that hay fever has increased over the last 30 years. On the other hand there does appear to have been an increase in allergic rhinitis in Switzerland over this period (Wuhtrich, 1991) The issue is very important because the answer potentially changes the argument about what kind of change we are observing. Has there been a general increase in allergic disease or a specific increase in asthma? Chronic sinusitis is a different issue because it is not simply an allergic disease. Although there is an eosinophil infiltrate in the sinus tissue, and eosinophilia in the peripheral blood, only 25-35% of the patients are allergic to common inhalant allergens (Newman et al, 1994; Hoover et al, 1997). Thus inhalant allergy to either pollens, airborne fungi or indoor allergens is not a major risk factor for sinusitis. It is possible that sinusitis is related to a T_H2 response to intrinsic allergens such as Dermatophytes or aerobic bacteria (Newman et al, 1994; Ward et al, 1989; Slunt et al, 1996). On the other hand, the apparent increase in chronic sinusitis has been observed over a period when the use of broad spectrum antibiotics has progressively increased.

If the increase in asthma was primarily caused by an increase in indoor allergen exposure, we would have to postulate that dust mite, animal dander and cockroach allergens had increased in parallel over a 35-year period. Second, we would need to explain why increases had continued for at least 15 years after changes in houses were complete. Third, we would have to explain why increases have occurred in age groups where indoor allergens are not the dominant risk factor, i.e. asthma in infants and adults. Thus as we approach the millennium, we are trying to understand what aspect of western society has either changed the immune response to allergens, increased the quantity of allergen that reaches the relevant part of the respiratory tract, or changed the physiological response of the lungs.

D. WHAT HAS CHANGED?

Prior to 1960 asthma was not considered to be a common disease, and even today, it remains rare in underdeveloped communities without electricity and motorized transport. In rural Africa and the villages of Papua New Guinea (PNG), the prevalence of asthma today is less than 1% (Yermaneberhan et al, 1997; Ng'anga, 1996; Woolcock & Peat, 1997). When Herxheimer and Prior studied Canadian Eskimo villages in the 1960's, they also found very few cases of asthma (Herxheimer & Schaefer, 1975). In 1997, in the village of Ohu, Madang Province, PNG, the children walk 2 miles to school, none of the children use inhalers and none complain of symptoms of "short wind." Thus we are trying to understand what there is in common between the change from rural to urban life and the changes that have occurred in almost all western societies over the last 35 years (Table II). In most countries these changes are not related to poverty, however asthma is very common and very severe among poor children living in the cities of North America. Most of the epidemiological data relates to school age children and in this age group, the strongest risk factor for asthma is sensitization to one (or more) of the common indoor allergens. In children it is possible to ask whether the change is: increased exposure to allergens; increased immune response; increase inflammatory response among allergic children or a change in the threshold for wheezing. The change we are concerned with has either directly effected inflammation in the lungs or acts selectively on allergic children.

1. Changes that could have increased exposure to allergens

Given that children and adults spend 23 hours plus per day indoors, it is not surprising that we are primarily concerned with house dust related allergens (Platts-Mills et al, 1992; Platts-Mills et al, 1997; Spengler & Sexton, 1983). As discussed earlier, the changes that occurred in houses in the UK, Australia and New Zealand were dramatic and could have improved both the 'nests' for mites and the conditions for mite growth. In other areas, the primary changes were in ventilation rate, with secondary effects on temperature. For example, in Sweden many houses changed from open fires to central heating. In other geographic areas, the changes in houses have not been impressive. Thus in central Atlanta windows are generally kept open and carpeting of any kind is unusual (Carter et al, 1997). Similarly, in Hong Kong, where asthma prevalence is also high, windows are generally open (Leung & Ho, 1997). In Atlanta and Hong Kong, it is easy to appreciate that mites grow well, but it is less easy to explain how houses have changed. Overall, it is difficult to identify a consistent change in houses that is common to all the areas where asthma has increased.

Table II
Changes in Western Society Over the Period 1960 - 1995

A. Increased exposure to indoor allergens.
- Increases in temperature, furnishings, and humidity, as well as changing cleaning techniques would improve the conditions for mite growth.
- Decreased ventilation rates and increased upholstered furniture could increase the quantities of airborne particulates, particularly animal dander.
- Increased time spent indoors.

B. Changes that could have influenced immune responsiveness.
- Diet - Increased calories, increased coloring and preservatives, increased sodium and changes in lipid profiles.
- Increased viral infections in early childhood.
- Broad spectrum antibiotics.

C. Outdoor air pollution including passive smoke.
- Classical air pollution: particulates and SO_2 has declined since 1960.
- Pollution related to vehicular traffic, oxides of nitrogen & diesel particulates.

D. Changes associated with the increase in sedentary visual entertainment.
- Prolonged periods spent without significant physical activity.
- Decline in physical activity: play or walking.

Increased time spent indoors, and perhaps more time spent sitting, may have increased exposure to dust related allergens, however this change is only of the order of 3 hours/day, i.e. from ~20 hours/day to ~23 hours/day (i.e. an increase of ~15%).

2. Changes that could have increased immune responsiveness

A relatively small shift in immune responsiveness of the population could produce a dramatic increase in the number of individuals presenting with a given disease (Seaton, 1994).

a. Diet: The possible role of dietary changes has been discussed for many years. In the 1970's, increased food coloring and preservative consumption was proposed as a cause of asthma. In a few communities decreased fish consumption could have increased immune responsiveness (Peat et al, 1994). A high sodium diet has also been associated with increased prevalence of wheezing (Burney, 1995). The shift from a high potassium diet to a high sodium diet in North America over the last 30 years, could have contributed to asthma prevalence particularly among African Americans. However, it is difficult to sort out a consistent pattern of dietary change that could have influenced the many different areas of the world where increases have been reported.

b. Infections: The role of viral infections as enhancers, and as precipitating factors for acute episodes of asthma, is now well accepted (Calhoun et al, 1994; Lemanske et al, 1989; Johnston et al, 1995; Duff et al, 1993). By contrast, the role of early viral infections in influencing the pattern of immune responses is much less clear. Two different models have been proposed:

Respiratory syncytial virus (RSV) infection in the first two years of life is a major cause of bronchiolitis and can induce a T_H2 response, eosinophil activation, and perhaps specific IgE antibody (Duff et al, 1993; Welliver et al, 1981; Srikiatkhachorn & Braciale, 1997). While the evidence about IgE ab to RSV remains unclear, there is excellent evidence that RSV proteins can induce eosinophil rich responses both in mice and humans (Srikiatkhachorn & Braciale, 1997; Rakes et al, 1999).

Early exposure to common respiratory viruses may also have a protective effect. This mechanism has been proposed as an explanation for the lower prevalence of asthma among second and subsequent children, as compared to oldest or only children (Lewis et al, 1996; Martinez, 1994). The argument is that an older child attending school or day care will bring viruses home to the younger siblings. It is not at all clear how this model would help to explain the very high prevalence of asthma among children living in overcrowded conditions in poor areas of American cities.

c. Antibiotics: Another change that could have altered immune responsiveness is the introduction of broad spectrum antibiotics. Penicillin was first used in the mid 1940's, but the widespread use of oral antibiotics started around 1960 and has increased steadily since then. The argument is that untreated bacterial infections stimulate macrophages to produce IL-12, which is a major stimulus to the differentiation of T_H1 cells (Zhou et al, 1995). If the infections are stopped or limited, we assume that the stimulus to T_H1 is decreased with a relative increase in T_H2 responses. Inherent in this argument is the view that the production of IL-12 would influence immune responses other than those that are stimulating the response (Fig. 3). The striking feature about antibiotic use is that these drugs were introduced at about the same time as the increase in asthma started and that their use has increased steadily since then. However, neither decreased infections, nor increased use of antibiotics, are a convincing explanation for the contrast between inner city America and low income populations in other parts of the world.

T Cell Differentiation During Human Immune Responses (Immune Deviation)

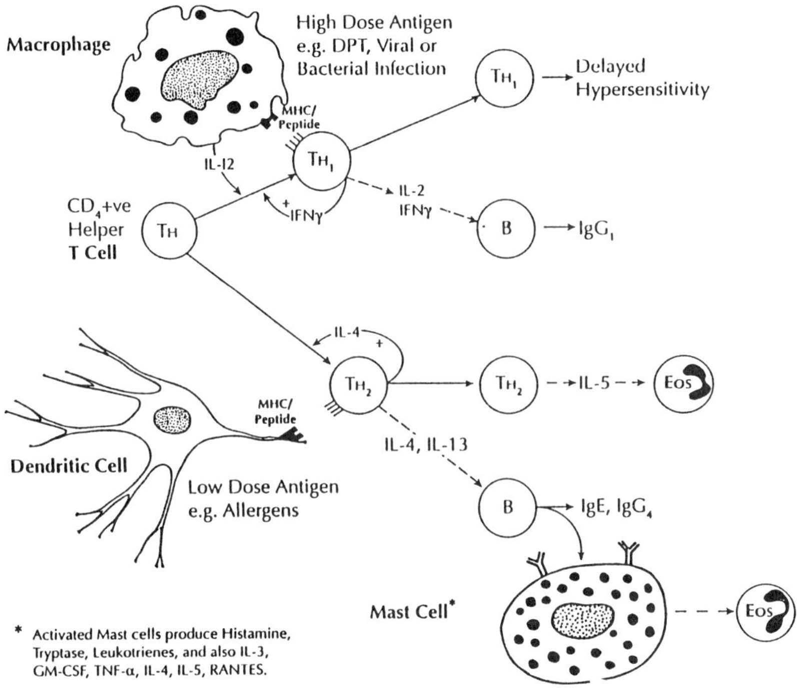

Figure 3. Immune responses to foreign proteins can come in many different forms but there is an important dichotomy between those in which the predominant T cells are T_H1 and those where they are predominantly T_H2; however, the nonallergic state is generally one of no response rather than being a T_H1 response. Thus, very few nonallergic individuals produce delayed hypersensitivity skin tests to common inhalant allergens such as dust mites, cat or pollens.

Overall there are many changes in Western society that could have changed immune responsiveness. Indeed, it is highly probable that changes in viral infections, diet and antibiotic use have each had an important effect on immune and inflammatory responses in the respiratory tract. However, it is difficult to see how any of these changes amount to a consistent effect on asthma prevalence of the size that has occurred in so many areas of the world.

3. Changes related to a sedentary life

In most societies prior to 1960, it was normal for children to play outdoors for 3 hours or more per day. In addition, walking or cycling several miles per day (e.g. to school) was a normal part of life. Over the last 30 years, physical activity among children has decreased dramatically. Indeed, the level of physical activity among children in the United States today would be astonishing to parents from 1960. One has to assume that the change of behavior among children has been accepted by parents because it has developed so insidiously. It is no surprise that there has been a progressive rise in obesity among children (USDHHS, 1996; Crater & Platts-Mills, 1998). What makes this more impressive is that this increase in the weight of children has occurred over a period when caloric consumption has apparently decreased (USDHHS, 1996). The decrease in physical activity has been achieved by the combination of motorized transport and indoor visual entertainment. The addictive effect of videos, television, and computers on young children is so profound that it is not uncommon for them to sit still for 4 hours or more per day (Gortmacher et al, 1996). The question is whether prolonged sitting is inherently harmful or whether the closely related decline in physical activity represents the loss of a protective effect on lung physiology. There are many children who do high levels of exercise and prolonged sitting. There are also very large numbers of children complaining of exercise-induced asthma. There is no simple relationship between asthma and obesity. What is certain is that there has been a progressive decline in the number of hours per week spent in physical activity over the last 30 years.

That full inspiration produces a decrease in airway resistance has been known for many years (Nadel & Tierney, 1961). Recently, Togias and his colleagues have demonstrated that prolonged tidal breathing, without full inspiration (or sighs), leads to a transient state of increased bronchial reactivity in normal individuals (Skloot et al, 1995). Thus it is possible that prolonged sedentary entertainment could contribute to bronchial reactivity. Certainly, sigh rates are decreased in normal individuals who are watching a video screen as compared to reading (Hark et al, 1998). At present it seems more likely that prolonged physical activity such as walking, playing or cycling has a beneficial effect. Even mild exercise could increase full inspirations to >500/hour. Several studies have found a beneficial effect of regular training on asthma (see Crater & Platts-Mills, 1998). Overall it is clear that there have been major changes in physical activity among children and adults over the period we are talking about. Furthermore, the decline in physical activity has been particularly profound among African-American children in large cities (partly due to the fear of violence on the streets) (Rosenstreich et al, 1997; Gergen & Weiss, 1990). At the same time asthma remains rare in those societies where houses lack electricity and walking is the normal form of transport (Yermaneberhan et al, 1997; Ng'anga, 1996).

CONCLUSIONS

We are studying a truly remarkable epidemic of wheezing which has been documented in many different developed countries. The disease is now accepted as an inflammatory disease, and the best accepted cause of this inflammation is inhalation of common indoor allergens. However, the increase in asthma has been reported in communities where the climates and the physical design of housing are completely different. Furthermore, the increase has been observed in all social classes. In keeping with this, many different sources of house dust related allergens have been found to be relevant.

In trying to understand why an allergic/ inflammatory disease of the lungs should have increased in prevalence by as much as 10-fold, it is necessary to consider all the changes that have occurred in western society over this time period. At present, none of the changes appears to explain the epidemic adequately. Thus neither increased exposure to allergens, dietary changes, alterations in bacterial or viral infections, or decreased physical activity can fully explain the epidemic. If so, it may be that it is the combination of two or more factors. It is interesting to compare asthma with some other diseases that are a feature of western society, i.e. hypertension, coronary arteriosclerosis and Type II diabetes. In each of these cases diet, exercise, and genetics have all been incriminated. It seems likely that asthma reflects a complex interaction between diet and genetics influencing the immune response to common environmental allergens. However, in addition the data from underdeveloped countries suggests that there may be profound protective effect of prolonged exercise which overrules the other factors. Thus, we may be able to identify a series of risk factors that are relevant to asthma in developed countries, but these would not be relevant in a community where walking or playing for two or more hours/day was normal behavior.

Sorting out the changes that are relevant to the increase in asthma is important from a public health and scientific point of view, but may also be relevant to treatment of individual cases. Current treatment is not adequate. The contrast between the dramatic effects of H2 blockers in reducing the need for peptic ulcer surgery, and the failure of inhaled steroids, or any other treatment, to control asthma sufficiently to prevent admissions to hospital, suggests that we have not yet fully understood this disease. Some of the changes that have occurred over the last 30 years may be irreversible. However, it would be very sad if we concluded that it was impossible to reverse the trend towards a population of unfit, overweight children, 10% of whom require inhaled steroids to keep their lungs from wheezing.

ACKNOWLEDGMENTS

Supported by NIH Grant Nos. AI-20565, AI-30840, and a grant from the Center for Indoor Air Research.

REFERENCES

Aalberse RC, van der Gaag R, van Leeuwen J. Serologic aspects of IgG4 antibodies. I. Prolonged immunization results in an IgG4-restricted response. *J of Immunol* 1983;130:722-726.

Aberg N. Asthma and allergic rhinitis in Swedish conscripts. *Clin Exp Allergy* 1989;19:59-63.

Altounyan RE. Changes in histamine and atropine responsiveness as a guide to diagnosis and evaluation of therapy in obstructive airways disease. In: Pepys J, Frankland AW, eds. *Disodium Chromoglycate in Allergic Airways Disease*. London: Butterworth, 1970:47-5.

Beasley R, Roche WR, Roberts JA and Holgate ST. Cellular events in the bronchi in mild asthma and after bronchial provocation. *Am Rev Resp Dis* 1989;139:806-817.

Bernton HS, McMahon TF, Brown H. Cockroach Asthma. *Brit J Dis Chest* 1972;66:61-66.

Broder I, Higgins, NW, Mathews KP, Keller JB. Epidemiology of asthma and allergic rhinitis in a total community. Tecumseh, Michigan. IV Natural history. *J Allergy* 1974;54:100.

Burney P. The origins of obstructive airways disease. A role for diet? *Am J Resp Crit Care Med* 1995;153:1292-1293.

Calhoun WJ, Dick EC, Schwartz LB, Busse WW. A common cold virus, rhinovirus 16, potentiates airway inflammation after segmental antigen bronchoprovocation in allergic subjects. *J Clin Invest* 1994;94:2200 -2208.

Call RS, Smith TF, Morris E, Chapman MD, Platts-Mills TAE. Risk factors for asthma in inner city children. *J Peds* 1992;121:862-866.

Carter M, Perzanowski M, Raymond A, Platts-Mills TAE. Sensitization and exposure of inner city children enrolled in a controlled trial of allergen avoidance for asthma. *J Allergy Clin Immunol* 1997;99:5410.

Chapman MD, Platts-Mills TAE. Purification and characterization of the major allergen from *Dermatophagoides pteronyssinus*-antigen P1. *J Immunol* 1980;125:587-592.

Cockcroft DW. The effect of 1 week's treatment with albuterol 200ng q.i.d. on immediate and late responses to allergen challenge. *J Allergy Clin Immunol* 1995;96:44.

Cooke RA, Van DerVeer A. Human sensitization. *J Immunol* 1916;1:20-

1-237.

Crater SE, Platts-Mills TAE. Searching for the cause of the increase in asthma. *Curr Opin Peds* 1998;10(6), in press.

Diaz-Sanchez D, Dotson RA, Takenaka H, Saxon A. Diesel exhaust particles induce local IgE production in vivo and alter the pattern of IgE messenger RNA isoforms. *J Clin Invest* 1994;94:1417-1425.

Duff AL, Pomeranz ES, Gelber LE, Price GW, Farris H, Hayden FG, Platts-Mills TAE, Heymann PW. Risk factors for acute wheezing in infants and children: viruses, passive smoke, and IgE antibodies to inhalant allergens. *Pediatrics* 1993;92:535-540.

Gelber LE, Seltzer LH, Bouzoukis JK, Pollart SM, Chapman MD, Platts-Mills TAE. Sensitization and exposure to indoor allergens as risk factors for asthma among patients presenting to hospital. *Amer Rev Resp Dis* 1993;147:573-578.

Gergen PJ, Weiss KB. Changing patterns of asthma hospitalization among children: 1979 to 1987. *JAMA* 1990;264:1688-1692.

Gleich GJ, Adolphson CR. The eosinophilic leukocyte: Structure and function. *Adv Immunol* 1986;39:177-253.

Gortmacher SL, Must A, Sobol AM, Peterson K, Colditz GA, Dietz WH. Television viewing as a cause of increasing obesity among children in the United States. *Arch Ped Adol Med* 1996;150:356-362.

Haahtela T, Lindholm H, Bjorksten F, Koskenvuo K, Laitenen LA. Prevalence of asthma in Finnish young men. *Brit Med J* 1990;301:2 66-268.

Hagy GW, Settipane GA. Prognosis of positive allergy skin tests in an asymptomatic population. A three year follow-up of college students. *J Allergy Clin Immunol* 1971;48:200.

Hark WT, Thompson WM, McLaughlin T, Platts-Mills TAE. The influence of behavior on resting sigh and respiratory rates in normal and asthmatic subjects. *J Allergy Clin Immunol* 1998;101:S183.

Herxheimer H, Schaefer O. Asthma in Canadian Eskimos. *N Eng J Med* 1975;291:1419-(Letter).

Hogg JC. Pathology of Asthma. *Asthma: Physiology, Immunopharmacology, and Treatment. Fourth International Symposium 1992*:17-25.

Hoover GE, Newman LJ, Platts-Mills TAE, Phillips CD, Gross CW, Wheatley LM. Chronic sinusitis: Risk factors for extensive disease. *J Allergy Clin Immunol* 1997;100:185-191.

Ingram JM, Sporik RB, Rose G, Honsinger R, Chapman MD, Platts-Mills TAE. Quantitative assessment of exposure to dog (Can f I) and cat (Fel d I) allergens: Relationship of sensitization and asthma among children living in Los Alamos, NM. *J Allergy Clin Immunol* 1995;96:449-56.

Jackson R, Sears MR, Beaglehole R and Rea HH. International trends in asthma mortality:1970 to 1985. *Chest* 1988; 94:914-919.

Johnston SL, Pattemore PK, Sanderson G, Smith S, Lampe F, Josephs L, Symington P, O'Toole S, Myint SH, Tyrrel DAJ, Holgate ST.

Community study of role of viral infections in exacerbations of asthma in 9-11 year old children. *Brit Med J* 1995;310:1225-1229.

Kang B, Vellody D, Homburger H, Yunginger JW. Cockroach cause of allergic asthma: Its specificity and immulogic profile. *J Allergy Clin Immunol* 1979;63:80-86.

Kern RA. Dust sensitization in bronchial asthma. *Med Clin N Am* 1921;5: 751-758.

Kerrebijn KF. Endogenous factors in childhood CNSLD: methodological aspects in population studies. In: Orie NGM, van der Lende R, eds. *Bronchitis III*. The Netherlands: Royal Vangorcum Assen, 1970:38-48.

Kuehr J, Frischer J, Meiner R. Mite exposure is a risk factor for the incidence of specific sensitization. *J Allergy Clin Immunol* 1994;94: 44-52.

Lemanske RF, Dick EC, Swenson CA, Vrtis RF, Busse WW. Rhinovirus upper respiratory infection increases airway hyperreactivity. *J Clin Invest* 1989;83:1-10.

Leung R, Ho P. 1994. Asthma, allergy and atopy in three southeast Asian populations. *Thorax* 1997;49:1205-1210.

Lewis S, Butland B, Strachan D et al. Study of the aetiology of wheezing illness at age 16 in two national British cohorts. *Thorax* 1996;51: 670-676.

Liu MC, Hubbard WC, Proud D, et al. Immediate and late inflammatory responses to ragweed antigen challenge of the peripheral airways in allergic asthmatics. *Amer Rev Respir Dis* 1991;144:51-58.

Lundback B. Asthma, chronic bronchitis and respiratory symptoms: prevalence and important determinants: *The obstructive lung disease in Northern Sweden Study 1*. Umea University 1993.

Martinez FD. Role of viral infections in the inception of asthma and allergies during childhood: could they be protective? *Thorax* 1994;49:1189-91.

Menon P, Rando RJ, Stankus RP, Salvaggio JE, Lehrer SB. Passive cigarette smoke-challenge studies: increase in bronchial hyperreactivity. *J Allergy Clin Immunol* 1992;89:560-566.

Michel O, Kips J, Duchateau J, Vertongen F, Robert L, Collet H, Pauwels R, Sergysels R. Severity of asthma is related to endotoxin in house dust. *Amer J Resp Crit Care Med* 1996;154:1641-6.

Mitchell N, Newacheck PW. Trends in the hospitalization for acute childhood asthma, 1970-1984. *Am J Public Health* 1986;76:1308-1311.

Molfino NA, Wright SC, Katz I, Tarlo S, Silverman F, McLean PA, Szalai JP, Raizenne M, Slutsky AS, Zamel N. Effect of low concentrations of ozone on inhaled allergen responses in asthmatic patients. *Lancet* 1991;338:199-203.

Nadel JA, Tierney DF. Effect of a previous deep inspiration on airway resistance in man. *J Appl Physiol* 1961;16:717-719.

Newman LJ, Platts-Mills TAE, Phillips CD, Hazen KC, Gross CW.

Chronic sinusitis. Relationship of computed tomographic findings to allergy, asthma and eosinophilia. *JAMA* 1994;271:363-367.

Ng'anga LW. The epidemiology of childhood asthma in Kenya. (Ph.D. thesis) McGill University, Montreal 1996.

Peat JK, van den Berg RH, Green WF, Mellis CM, Leeder SR, Woolcock AG. Changing prevalence of asthma in Australian children. *Brit Med J* 1994;308:1591-1596.

Peat JK, Tovey E, Toelle BG, Haby MM, Gray EJ, Mahmic A, Woolcock AJ. House dust mite allergens: A major risk factor for childhood asthma in Australia. *Am J Respir Crit Care Med* 1996;153:141-146.

Peden DB, Setzer RW, Jr., Devlin RB. Ozone exposure has both a priming effect on allergen-induced responses and an intrinsic inflammatory action in the nasal airways of perennially allergic asthmatics. *Am J Respir Crit Care Med* 1995;151(5):1336-45.

Perzanowski M, Ronmark E, Lundback B, Platts-Mills TAE. Relevance of allergens from cats and dogs to asthma in the northern most province of Sweden: Schools as a major souce of exposure. *J Allergy Clin Immunol* 1999, in press.

Platts-Mills TAE, Tovey ER, Mitchell EB, Moszoro H, Nock P, Wilkins SR. Reduction of bronchial hyperreactivity during prolonged allergen avoidance. *Lancet* 1982;2:675-678.

Platts-Mills TAE, Mitchell EB, Tovey ER, Chapman MD, Wilkins SR. Airborne allergen exposure, allergen avoidance and bronchial hyperreactivity. In: Kay AB, Austen KF, Lichtenstein LM, eds. *Asthma: Physiology, Immunopharmacology and Treatment, Third International Symposium*. London: Academic Press, Inc., 1984:297-314.

Platts-Mills TAE, Thomas WR, Aalberse RC, Vervloet D, Chapman MD, et al, (Co-Chairmen). Dust mite allergens and asthma: Report of a Second International Workshop. *J Allergy Clin Immunol* 1992; 89:1046-1060.

Platts-Mills TAE, Sporik RB, Wheatley L, Heymann PW. Is there a dose response relationship between exposure to indoor allergens and symptoms of asthma [editorial]. *J Allergy Clin Immunol* 1995;96: 435-440.

Platts-Mills TAE, Sporik RB, Chapman MD, Heymann PW. The role of domestic allergens. In: Chadwick DJ, Cardew G, eds. *Rising Trends in Allergy*. Chichester, UK; John Wiley & Sons Ltd., 1997;173-189.

Platts-Mills TAE, Vervloet D, Thomas WR, Aalberse RC, Chapman MD (Co-Chairmen). Indoor allergens and asthma. Third International Workshop, Cuenca Spain. *J Allergy Clin Immunol* 1997;100:S1-S24.

Rakes GP, Arruda E, Ingram JM, Hoover GE, Zambrano J, Hayden FG, Platts-Mills TAE, Heymann PW. Rhinovirus and respiratory syncytial virus in wheezing children requiring emergency care: IgE and eosinophil analysis. *Am J Resp Crit Care Med*, Feb. 1999.

Rando RJ, Menon PK, Poovey HG, Lehrer SB. Assessment of multiple markers of environmental tobacco smoke (ETS) in controlled, steady-state atmospheres in a dynamic test chamber. *Am Ind Hyg Assoc J* 1992;53:699-704.

Ronmark E, Lundback B, Jonsson E, Platts-Mills TAE. Asthma, type-1 allergy and related conditions in 7- and 8-year-old children in Northern Sweden: prevalence rates and risk factor pattern. *Resp Med* 1998; 92:316-324.

Rosenstreich DL, Eggleston P, Kattan M, Baker D, Slavin RG, Gergen P, Mitchell H, McNiff-Mortimer K, Lynn H, Ownby D, Malveaux F. The role of cockroach allergy and exposure to cockroach allergen in causing morbidity among inner-city children with asthma. *N Engl J Med* 1997;336:1356-1363.

Sears MR, Taylor DR, Print CG, et al. Regular inhaled beta-agonist treatment in bronchial asthma. *Lancet* 1990;336:1391-1396.

Seaton A, Godden DJ, Brown K. Increase in asthma: a more toxic environment or a more susceptible population? *Thorax* 1994;49:171-174.

Shaver JR, Zangrilli JG, Cho SK, Cirelli RA, Pollice M, Hastie AT, Fish JE, Peters SP. Kinetics of the development and recovery of the lung from IgE-mediated inflammation. *Amer J Res Crit Care Med* 1997;155:442-448.

Skloot G, Permutt S, Togias A. Airway hyperresponsiveness in asthma. *J Clin Invest* 1995;96:2393-2403.

Slunt JB, Taketomi EA, Woodfolk JA, Hayden ML, Platts-Mills TAE. The immune response to Trichophyton tonsurans: Distinct T cell cytokine profiles to a single protein among subjects with immediate and delayed hypersensitivity. *J Immunol* 1996:5129-5197.

Smith JM, Disney ME, Williams JD, Goels ZA. Clinical significance of skin reactions to mite extracts in children with asthma. *Brit Med J* 1969;1:723-726.

Spengler JD, Sexton K. Indoor air pollution: a public health perspective. *Science* 1983;221:9-17.

Spieksma FT, Voorhorst R. Comparison of skin reactions to extracts of house dust, mites, and human skin scales. *Acta Allergol* 1969; 24:124-146.

Sporik RB, Holgate ST, Platts-Mills, TAE, Cogswell J. Exposure to house dust mite allergen (Der p I) and the development of asthma in childhood: A prospective study. *N Eng J Med* 1990;323:502-507.

Sporik RB, Chapman MD, Platts-Mills TAE. House dust mite exposure as a cause of asthma (Editorial). *Clin Exp Allergy* 1992;22:897-906.

Sporik R, Ingram JM, Price W, Sussman JH, Honsinger RW, Platts-Mills TAE. Association of asthma with serum IgE and skin-test reactivity to allergens among children living at high altitude: tickling the dragon's breath. *Amer J Res Crit Care Med* 1995;151:1388-1392.

Squillace SP, Sporik RB, Rakes G, Couture N, Lawrence A, Merriam S, Zhang J, Platts-Mills TAE. Sensitization to dust mites as a dominant risk factor for adolescent asthma: Multiple regression analysis of a population-based study. *Amer J Resp Crit Care Med* 1997; 156:1760-1764.

Srikiatkhachorn A, Braciale TJ. Virus-specific CD8+ T lymphocytes downregulate T helper cell type 2 cytokine secretion and pulmonary eosinophilia during experimental murine respiratory syncytial virus infection. *J Exp Med* 1997;186:421-432.

Tattersfield AE. Use of beta 2 agonists in asthma: much ado about nothing? Still cause for concern. *Brit Med J* 1994;309:795-5.

Tovey ER, Chapman MD, Platts-Mills, TAE. Mite faeces are a major source of house dust allergens. *Nature* 1981;289:592-593.

Tovey ER, Chapman MD, Wells CW, Platts-Mills TAE. The distribution of dust mite allergen in the houses of patients with asthma. *Amer Rev Resp Dis* 1981;124:630-635.

U.S. Dept of Health and Human Services. Physical Activity and Health: Report of the Surgeon General. Atlanta: U.S. Dept. of Health and Human Services, Centers for Disease Control and Prevention, National Center for Chronic Disease Prevention and Health Promotion, International Medical Publishing, 1996: 175-206.

Voorhorst R, Spieksma FThM, Varekamp N. *House dust mite atopy and the house dust mite Dermatophagoides pteronyssinus* (Troussart, 1897). Leiden: Stafleu's Scientific Publishing Co., 1969.

Ward GW, Jr., Karlsson G, Rose G, Platts-Mills TAE. Tricophyton asthma: sensitization of bronchi and upper airways to dematophyte antigen. *Lancet* 1989;1:859-862.

Welliver RC, Wong DT, Sun M, Middleton E, Vaughan RS, Ogra PL. The development of respiratory synctial virus-specific IgE and the release of histamine in nasopharyngeal secretions after infection. *N Eng J Med* 1981;305:841-846.

Wierenga EA, Snoek M, DeGroot C, Chretien I, Bos JD, Jansen HM, Kapsenberg ML. Evidence for compartmentalizationof functional subsets of CD4+ lymphocytes in atopic patients. *J Immunol* 1990;144: 4651-4656.

Woolcock AJ, Peat JK. Evidence for the increase in asthma worldwide. In: Ciba Foundation, ed. *The Rising Trends in Asthma*. Chichester: John Wiley & Sons, 1997:122-139.

Wuhtrich B. In Switzerland pollinoisis has really increased in the last decade. *ACI News* 1991;3:41-44.

Yermaneberhan H, Bekele Z, Venn A, Lewis S, Parry E, Britton J. Prevalence of wheeze and asthma and relation to atopy in urban and rural Ethiopia. *The Lancet* 1997;350:85-90.

Zhou P, Sieve MC, Bennett J, Kwon-Chung KJ, Tewari RP, Gazzinelli RT, Sher A, Seder RA. IL-12 prevents mortality in mice infected with Histoplasma capsulatum through induction of IFN-gamma. *J Immunol* 1995;155:785-795.

Chapter 2

VIRAL MODULATION OF AIRWAY INFLAMMATION

Scherer P. Sanders, Ph.D. and David Proud, Ph.D.

I. INTRODUCTION

Asthma is an inflammatory and obstructive pulmonary disease that affects approximately 12 million Americans (Weiss et al., 1992). Despite improved knowledge of disease pathogenesis and the availability of new therapies, the prevalence of, and the mortality and morbidity associated with, asthma has increased steadily during the past 15 years (Evans et al., 1987; Halfon and Newacheck, 1986; Sly, 1988; Weiss et al., 1992). For 1990, the cost of illness associated with asthma in the United States has been estimated to be approximately $6 billion, with the treatment of acute exacerbations accounting for about half of this amount (Smith et al., 1997; Weiss et al., 1992). Clearly, therefore, a better understanding of the mechanisms responsible for triggering asthma attacks could have a major impact on health care costs.

Although several precipitating factors, including exercise or hyperventilation, and exposure to noxious gases, drugs, or allergens can lead to acute aggravations of asthma, this chapter will focus upon the potential mechanisms by which upper respiratory viral infections (URI), arguably the most common risk factor associated with asthma attacks, induce asthma exacerbations.

Viral infections have been implicated as a major factor both in the development of asthma and in the induction of asthma exacerbations. In infancy, viral infections provoke episodes of wheezing, which may be recurrent. Respiratory syncytial virus is the primary virus detected in children under the age of 2 who present with wheezing in an emergency room setting (Duff et al., 1993). It has been shown, however, that most infants who wheeze during respiratory infections have less difficulty as they grow older, leading some to conclude that URI cannot be the sole cause for the development of asthma (Martinez et al., 1995). Although the role of viruses in the development of asthma is unclear, there is strong evidence linking URI to asthma exacerbations.

The association between viral infection and acute episodes of asthma was first noted during influenza epidemics in 1957 and 1958 (Podosin and Felton, 1958; Rebhan, 1957). At that time, it was assumed that bacterial infections were also a major risk factor for asthma attacks (Feingold, 1959). In subsequent prospective studies, however, it was shown that, while up to 40% of asthma exacerbations were associated with the presence of a

1-56670-373-5/99/$0.00+$.50

detectable viral infection, no such relationship was observed for bacterial infection. Indeed, bacterial infections were detected in the respiratory tract of asthmatics with equal incidence during both asymptomatic and symptomatic periods (Hudgel et al., 1979; McIntosh et al., 1973). By contrast, numerous subsequent studies detected viruses in airway secretions in up to 40% of exacerbations of asthma in children and in up to 20% of exacerbations in adults (Pattemore et al., 1992), while the detection rate during asymptomatic periods in both asthmatics and nonasthmatics was around 3% (Horn et al., 1979; Hudgel et al., 1979; Jennings et al., 1987). Moreover, it is apparent that the rates of URI detected during asthma attacks in early studies were low because human rhinoviruses and coronaviruses are difficult to detect by standard culture methods (Larson et al., 1980). In prospective studies using reverse transcription-polymerase chain reaction (RT-PCR), in addition to conventional virology, viral pathogens were detected in greater than 80% of asthma attacks in children ages 9 to 11, with rhinoviruses being associated with almost 60% of exacerbations (Johnston et al., 1995). In adults, 80% of subjective asthma exacerbations, and 44% of episodes of peak flow deteriorations, were associated with URI, with 64% of episodes again being associated with rhinovirus infection (Nicholson et al., 1993). By contrast, the detection rates of HRV in apparently healthy subjects using RT-PCR was approximately 2% (Nicholson et al., 1993). Rhinoviruses were also noted to be the most common viral pathogen associated with asthma attacks in young children over 2 years old presenting in an emergency room setting (Duff et al., 1993).

Epidemiologic studies also demonstrate a clear temporal relationship between outbreaks of viral infections and asthma exacerbations (Beasley et al., 1988; Johnston et al., 1996; McIntosh et al., 1973; Mitchell et al., 1976). In pediatric populations, hospital admissions for asthma are less common during school holidays but peak during the early part of school terms; data that has been interpreted to suggest that rapid spreading of viral infections in the school environment is a risk factor for asthma attacks (Johnston et al., 1996; Storr and Lenney, 1989). There have been few direct comparisons of the importance of viral infections and aeroallergens as triggering agents for asthma exacerbations, but two such studies suggested little relationship between the seasonal distribution of asthma attacks and changes in pollen and spore counts, or known allergen exposure, while confirming strong associations between asthma exacerbations and the presence of URIs (Carlsen et al., 1984; Potter et al., 1984). One recent study, however, did demonstrate increased hospital admissions during months associated with peak levels of house dust mite (Johnston et al., 1996).

Given the predominant association between rhinovirus infections and asthma exacerbations in humans from age 2 to adulthood (Duff et al., 1993;

Johnston et al., 1995; Nicholson et al., 1993), we have elected to use rhinovirus for our studies seeking to understand how viral infections may trigger asthma attacks.

II. RHINOVIRUS INFECTIONS AND LOWER AIRWAYS FUNCTION

The role of rhinovirus infections in asthma attacks is supported further by studies of lower airways physiology. Although early studies of naturally acquired, uncharacterized URI showed increased airways reactivity in normal subjects (Aquilina et al., 1980; Empey et al., 1976), no such effect was observed in an additional study of naturally acquired colds (Jenkins and Breslin, 1984) or in more controlled studies using experimental rhinovirus infections (Angelini et al., 1997; Fraenkel et al., 1995; Gern et al., 1997; Halperin et al., 1983). By contrast, there is much stronger evidence showing that rhinovirus infection increases airways reactivity in allergic and asthmatic subjects. Although one study found that only a subpopulation of asthmatic subjects showed increased reactivity (Halperin et al., 1985), several studies have shown significant shifts for populations of both asthmatic (Cheung et al., 1995; Fraenkel et al., 1995; Grünberg et al., 1996) and allergic subjects (Gern et al., 1997; Lemanske et al., 1989). These changes in hyperreactivity are usually seen beginning about 2 or 3 days after viral exposure and have been reported, in some instances, to last for days to weeks after infection (Cheung et al., 1995; Lemanske et al., 1989). In atopic subjects, it has also been reported that rhinovirus infections may exacerbate responses to subsequent allergen provocation. Not only does rhinovirus infection increase the likelihood that a subject will demonstrate a late phase response to allergen challenge (Lemanske et al., 1989), but viral infection also results in higher concentrations of plasma histamine after allergen challenge (Calhoun et al., 1991), as well as increased histamine levels in BAL fluid 48 hours after allergen challenge (Calhoun et al., 1994).

Although there is precedent for rhinovirus infection to decrease airway function in asthmatic and atopic subjects (Halperin et al., 1985; Trigg et al., 1996), such changes have been relatively modest and have not been observed in all subjects. It is reasonable to argue that, if rhinoviruses are a major risk factor for asthma attacks, experimental viral infection routinely should induce significant decreases in airway function. It must be remembered, however, that, for safety reasons, the majority of experimental infections have used either atopic, nonasthmatic subjects, or subjects with very mild asthma. Other factors that may also compound this issue are the use of a limited number of safety tested rhinovirus strains that may have lost some virulence with passage; the use of relatively low doses of challenge virus in some studies, and, in some instances, the use of subjects with some preexisting neutralizing serum antibodies to the viral strain. Further studies are required, therefore, to resolve this issue.

III. PATHOGENESIS OF RHINOVIRUS INFECTIONS

Assuming that rhinoviruses are a precipitating factor for asthma exacerbations, any consideration of the mechanisms by which exacerbations are induced should be framed within the context of our current understanding of the pathogenesis of rhinovirus infections.

Rhinoviruses are small icosahedral, ribonucleic acid-containing viruses that belong to the family of picornaviruses. The viral genome consists of a single strand of positive-sense RNA (Rueckert, 1996). There are currently 102 identified serotypes of rhinovirus. Approximately 90% of these serotypes belong to the so-called major group of rhinoviruses, which share intercellular adhesion molecule-1 (ICAM-1) as a cell surface receptor (Greve et al., 1989; Staunton et al., 1989). Members of the low density lipoprotein receptor family have been shown to serve as receptors for the members of the minor group of rhinoviruses (Hofer et al., 1994). Rhinoviruses are the most commonly isolated viruses from subjects experiencing common colds (Couch, 1996; Gwaltney et al., 1966). Despite numerous studies of both naturally acquired and experimentally induced rhinovirus infections, our understanding of disease pathogenesis is far from complete, but several important observations have been made. In contrast to viruses, such as influenza or respiratory syncytial virus, which could conceivably cause symptoms via their direct cytotoxic effects, rhinoviruses cause no marked changes in the morphology or integrity of the airway mucosa (Winther et al., 1984a; 1984b). Rhinovirus infections are associated, however, with the generation of proinflammatory mediators. The potent proinflammatory peptides, bradykinin and lysylbradykinin have been detected in increased amounts in airway secretions of symptomatic subjects with both experimentally induced and naturally acquired rhinovirus colds (Naclerio et al., 1988; Proud et al., 1990). In experimental rhinovirus infections, increased kinin generation was first seen at the same time that symptoms were first recorded, and this temporal relationship was maintained during the symptomatic phase of the cold. Moreover, in infected subjects, absolute kinin levels correlated with symptom scores (Naclerio et al., 1988). By contrast, rhinovirus infections do not result in mast cell activation, since no increases in histamine were detected in either experimentally induced or naturally acquired rhinovirus infections (Naclerio et al., 1988; Proud et al., 1990). Increased production of certain cytokines, including IL-1 (Proud et al., 1994), IL-6 (Zhu et al., 1996) and IL-8 (Grünberg et al., 1996; Zhu et al., 1997) also occurs during symptomatic rhinovirus infections. The infiltration of selected inflammatory cell populations also correlates temporally with the symptomatic response to rhinovirus infections. Infection of nonatopic subjects, or of seasonal atopics out of season, results in increased numbers of neutrophils both in the nasal mucosa (Winther et al., 1984b), and in recovered nasal secretions (Levandowski et al., 1988;

Naclerio et al., 1988), as well as in an acute decrease in total circulating lymphocytes (Levandowski et al., 1986) and an increase of lymphocytes in nasal secretions (Levandowski et al., 1988).

Rhinovirus infection also has been shown to enhance lower airway inflammation. Infection increased recruitment of eosinophils into BAL 48 hours after allergen challenge of allergic, but not nonallergic subjects; a finding that persisted for a month after infection (Calhoun et al., 1994). Rhinovirus induced exacerbation of lower airway inflammation also has been observed, without allergen provocation, in asthmatic, but not normal, subjects. Rhinovirus infection increased the numbers both of submucosal lymphocytes and of epithelial eosinophils. Again, the eosinophilia persisted into convalescence (Fraenkel et al., 1995). The difference in cellular recruitment between atopic and nonatopic subjects may be one important feature in the strong relationship noted between allergic status and viral induction of asthma attacks (Busse and Gern, 1997; Duff et al., 1993), and will be discussed in more detail below.

The respiratory epithelial cell is the primary site of rhinovirus infection. During *in vivo* infections, rhinovirus has been detected in nasal epithelial cells, but not other cell types, by immunofluorescence (Turner et al., 1982), and by *in situ* hybridization (Arruda et al., 1995; Bardin et al., 1994). Virus is detected in only a limited number of epithelial cells using these techniques, raising the possibility that there are foci of infection, rather than a uniform viral spreading throughout the airways. In the lower airways, it has proven difficult to culture rhinoviruses from bronchoalveolar lavage fluids obtained from infected individuals (Calhoun et al., 1994; Halperin et al., 1985), but rhinovirus RNA has been detected by RT-PCR in lower airway cells from subjects after experimental infection (Gern et al., 1997).

At this point, the sequence of events leading from viral infection to the manifestation of symptoms is unknown. A reasonable hypothesis, however, is that rhinovirus infection alters epithelial cell function in a manner that leads to the enhancement of lower airways inflammation, and that this enhanced inflammation is responsible for asthma exacerbations.

IV. RHINOVIRUS INDUCTION OF EPITHELIAL CYTOKINES

The epithelial cell can play an active role in airway inflammation by releasing a variety of proinflammatory mediators, including eicosanoids, peptides, reactive oxygen species and a broad range of cytokines (Polito and Proud, 1997). The hypothesis that viral infection modifies epithelial cell function is supported by *in vitro* studies showing that infection of epithelial cells with rhinovirus, induces the production of cytokines, including IL-1, IL-8, IL-6, IL-11 and GM-CSF (Einarsson et al., 1996; Sanders et al., 1998; Subauste et al., 1995; Terajima et al., 1997; Zhu et al., 1997; Zhu et al., 1996). More recent studies from our laboratory have extended these findings to show that epithelial cell production of the chemoattractant cytokines, eotaxin and RANTES is also induced upon rhinovirus infection. Many of the experiments on cytokine generation use crude viral preparations that have been propagated through diploid cells. Nonetheless, generation of cytokines is a specific viral effect and not due to contaminants in viral stocks because cytokine generation is largely prevented if the virus is inactivated by exposure to ultraviolet light, and because purified viral preparations also generate cytokines (Figure 1). The potential role of epithelial cytokine generation in the modulation of airway function is supported by the fact that respiratory syncytial virus and influenza, viruses that are also associated with asthma exacerbations, induce epithelial production of a similar spectrum of cytokines (Arnold et al., 1994; Becker et al., 1993; Choi and Jacoby, 1992; Einarsson et al., 1996; Matsukura et al., 1996; Noah and Becker, 1993; Saito et al., 1997). Moreover, increased concentrations of many of these cytokines have been detected in nasal secretions during experimentally induced, or naturally acquired URI (Grünberg et al., 1996; Noah et al., 1995; Proud et al., 1994; Teran et al., 1997; Zhu et al., 1997; Zhu et al., 1996).

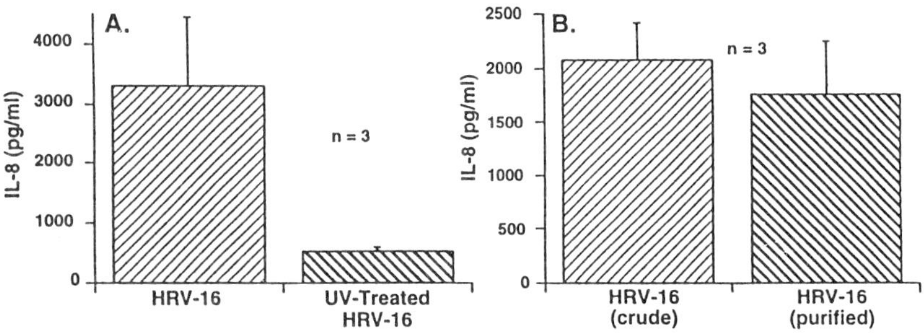

Figure 1: Specificity of rhinovirus induction of epithelial cytokine production. Panel A; rhinovirus induction of IL-8 is inhibited by inactivation of the virus by treatment with ultraviolet light. Panel B; Similar levels of cytokine production are seen using the identical titers of a crude virus preparation and of the same viral preparation after purification by sucrose density gradient centrifugation.

The molecular mechanisms by which rhinovirus infection induces cytokine production from epithelial cells have only recently begun to be examined. It has been shown that rhinovirus increases steady state mRNA expression for some cytokines, most notably IL-8 and IL-6 (Sanders et al., 1998; Zhu et al., 1997; Zhu et al., 1996). This seems to be due to an increased rate of transcription and is mediated, at least in part by the activation of the transcription factor, NF-κB (Zhu et al., 1997; Zhu et al., 1996). Few studies have focused on relating cytokine production to viral replication. It seems, however, that the generation of IL-8 and IL-6 occurs early in the process of viral infection and is not dependent upon viral replication *per se*, since IL-8 and IL-6 production is increased transiently, while viral replication continues unabated from several additional days (Sanders et al., 1998; Subauste et al., 1995). The relationship of the generation of other epithelial cytokines to rhinovirus replication has not yet been examined.

The production of epithelial-derived cytokines could play an important role in the pathogenesis of viral infections and, in some cases, asthma exacerbations. For example, targeted overexpression of IL-11 in murine airways has also been shown to be associated with lymphocytic infiltration, airways remodeling and bronchial obstruction (Tang et al., 1996). With regard to other epithelial cytokines, IL-8 is chemotactic for neutrophils and some T lymphocytes and is a direct activator of neutrophils (Larsen et al., 1989; Leonard and Yoshimura, 1990; Thelen et al., 1988), while IL-6 stimulates the proliferation and activation of T lymphocytes (Haussiau et al., 1989; Holsti and Raulet, 1989), and epithelial overexpression of IL-6 in transgenic mice leads to a marked lymphocytic infiltration of the airways (DiCosmo et al., 1994). Thus, IL-8 and IL-6 could play a role in the neutrophil and lymphocyte recruitment that is characteristic of rhinovirus infection in all subjects. It also has been suggested that IL-8 contributes to increased lower airway neutrophilia observed during some asthma exacerbations (Fahy et al., 1995). A particularly intriguing hypothesis, however, is that virally-induced epithelial production of RANTES, eotaxin and GM-CSF plays a central role in inducing asthma attacks by recruiting and activating airway eosinophils. RANTES is chemotactic for eosinophils, memory T lymphocytes, monocytes and basophils (Rot et al., 1992; Schall et al., 1990). Increased amounts of RANTES have been detected in the BAL fluids of asthmatic subjects (Alam et al., 1996) and RANTES has been suggested to be among the most important eosinophil chemoattractants in the asthmatic lung (Venge et al., 1996). Rhinovirus infection also causes epithelial cells to release eotaxin, a chemokine that selectively recruits eosinophils to sites of inflammation and is not chemotactic for monocytes, lymphocytes or neutrophils (Garcia-Zepeda et al., 1996; Ponath et al., 1996). GM-CSF promotes the survival, and enhances the activation, of eosinophils (Lopez et al., 1986; Owen et al., 1987). Activated eosinophils are present in the upper airways of wheezing children presenting in the emergency room (Duff et al., 1993). Moreover, as noted above, rhinovirus infection increases recruitment of eosinophils into bronchoalveolar lavage fluids 48 hours after allergen challenge of allergic subjects (Calhoun et al., 1994). Rhinovirus-induced increases in submucosal lymphocytes and in epithelial eosinophils also have been observed, without allergen provocation, in asthmatic, but not normal, subjects (Fraenkel et al., 1995). Eosinophils are considered to be important effector cells in asthma, releasing toxic basic proteins, as well as bronchospastic mediators, such a leukotrienes, (Gleich,

1990); and the presence of eosinophils in the bronchial mucosa is a cardinal feature of the disease (Djukanovic et al., 1990). Eosinophil numbers not only increase during asthma exacerbations (Laitinen et al., 1991) but correlate with the degree of airway hyperresponsiveness and with disease severity (Bousquet et al., 1990; Laitinen et al., 1991; Wardlaw et al., 1988). Therefore, increased bronchial eosinophilia during rhinovirus infection provides a potential mechanism for enhanced reactivity, airflow obstruction and disease exacerbation. If this hypothesis is accepted, however, important issues that must be considered include why viral infection leads to bronchial eosinophilia only in asthmatic or atopic subjects, and why viral infection does not always cause an asthma attack in every asthmatic.

V. SELECTIVE EOSINOPHIL RECRUITMENT

There are several potential explanations for why bronchial eosinophilia is observed *in vivo* after rhinovirus infection in asthmatics and atopics, but not in normals. One possibility would be if spreading of rhinovirus to lower airway epithelial cells occurs in atopics but not in normals. Rhinovirus has been detected in bronchial epithelial cells of infected atopic subjects (Gern et al., 1997). Although it seems unlikely that the ability of rhinovirus to reach the lower airways varies with atopic status, no studies have been performed looking for rhinovirus in the lower airways of infected normal subjects. A second possibility is that viral infection induces epithelial cytokine production only in atopic subjects. Again, this seems unlikely, since rhinovirus infection induces epithelial cytokine production *in vitro* without the need for cells to be derived from atopic donors, and IL-8 has been shown to be generated in nasal epithelium of normal subjects upon viral infection (Becker et al., 1993). Perhaps the most likely hypothesis to explain selective eosinophil recruitment in atopic subjects, however, is that rhinovirus induces epithelial cytokine production in all subjects, but that only eosinophils from atopic or asthmatic subjects respond to RANTES and eotaxin because they have been "primed" by *in vivo* exposure to cytokines such as IL-5, IL-3 or GM-CSF.

The concept of priming is by no means new. Increased chemotactic and chemokinetic responses have been observed with eosinophils obtained from patients with asthma, compared to those from nonasthmatic subjects, suggesting an enhanced activation state of eosinophils in asthmatics (Hakansson et al., 1990; Warringa et al., 1993). It also has been reported that eosinophils obtained during the birch pollen season from subjects with birch pollen allergy show enhanced expression of the integrins CD18 and CD49d, and increased adhesion to VCAM-1 and ICAM-1, compared to eosinophils obtained from the same subjects before the pollen season (Hakansson et al., 1997). It is known that IL-5, IL-3 and GM-CSF are involved in eosinophil activation in allergic diseases and augment a broad range of functional responses of these cells (Borish and Rosenwasser, 1996), including enhanced adhesion and migration (Bochner and Schleimer, 1994). Incubation of eosinophils from mildly atopic donors with IL-5 enhances *in vitro* transendothelial migration to RANTES (Ebisawa et al., 1994).

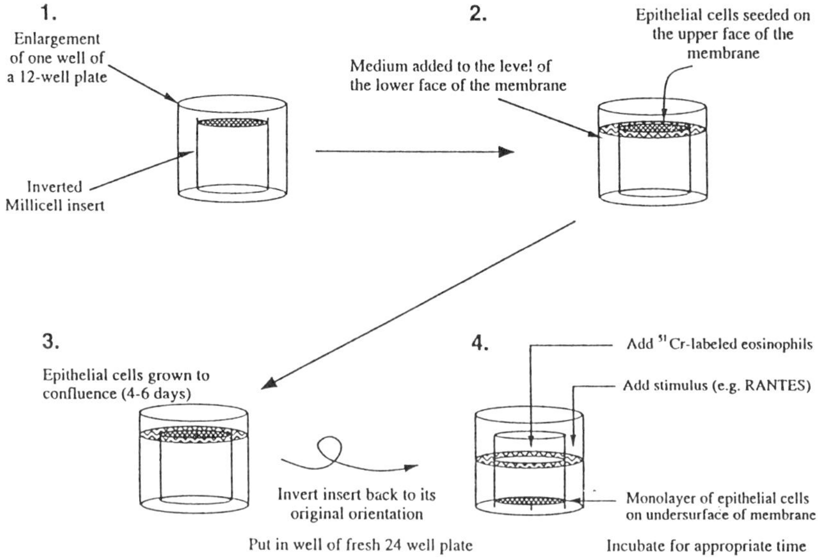

Figure 2: Model for the study of eosinophil migration across an epithelial layer in the physiologically relevant, basolateral-to-luminal direction.

Our own studies in this area have focused on examining the effects of priming on transepithelial migration of eosinophils in response to chemokines. In contrast to transendothelial migration, which has been intensively studied, little is known about the process governing the movement of eosinophils through the epithelial barrier. To examine this process it was necessary to first develop a model system in which migration of eosinophils across the epithelium could be studied in the physiologically relevant, basolateral-to-luminal direction. To accomplish this, Millicell culture well inserts (12 mm diameter polycarbonate membrane with 3.0 μm pores) are inverted in individual wells of 12-well culture plates filled with medium to the lower face of the membrane. Epithelial cells are seeded on the upper face and are fed from below until confluence is achieved (4-6 days). Confluence of the epithelial cell monolayers is confirmed by light microscopy and by Wright's staining of representative membranes. For the transepithelial migration assays, the culture well inserts are returned to their original orientation in 24 well plates (Figure 2). Purified peripheral blood eosinophils that have been radiolabeled by uptake of ^{51}Cr are added to the upper well, while the relevant chemotactic stimulus is added to the lower well. The volumes of fluids used in the upper and lower wells are selected to avoid any hydrostatic pressure gradient. After an appropriate incubation time, the inserts are discarded, and eosinophils that have migrated into the outer well are lysed by the addition of ammonium hydroxide. The level of radioactivity associated with the ^{51}Cr in each sample is determined in a gamma counter, and the percent migration is calculated. Our results with this model not only show that IL-5 treatment enhances transepithelial migration of eosinophils from atopic subjects in response to RANTES but, most importantly, that eosinophils from a nonatopic donor undergo transepithelial migration to RANTES *only* if first exposed to IL-5 (Figure 3).

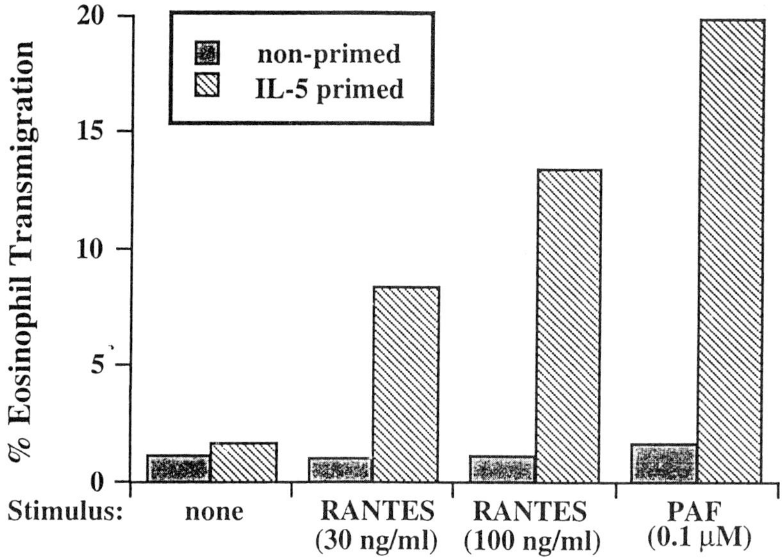

Figure 3: Transepithelial migration of eosinophils from a nonatopic donor in response to RANTES. The solid bars show that no migration occurs using freshly isolated eosinophils, while hatched bars show marked migration after incubation with IL-5 (5 ng/ml).

The need for "primed" eosinophils to respond to rhinovirus-induced epithelial chemokines may explain the strong relationship noted between viral induction of asthma attacks and atopic status (Busse and Gern, 1997; Duff et al., 1993). If this concept is valid, it may be asked why nasal eosinophilia has not been observed in experimental rhinovirus infection studies. This is likely because such infection protocols traditionally use normal, or asymptomatic atopic, subjects in order to avoid symptom scores being affected by allergy symptoms (Farr et al., 1990; Naclerio et al., 1988; Proud et al., 1994; Skoner et al., 1995). It is in these populations that neutrophil and lymphocyte infiltration is observed (Levandowski et al., 1988; Naclerio et al., 1988; Winther et al., 1984). It may be predicted, therefore, that rhinovirus infection of symptomatic atopic subjects should enhance nasal eosinophilia, but, to date, no such study has been performed.

VI. ANTIVIRAL EFFECTS OF NITRIC OXIDE

Several explanations could be presented as to why viral infection does not always induce an asthma attack in asthmatic subjects. These include variations in viral load, and variations in any of the multiple aspects of the host antiviral and immune response. It has recently become clear that one important component of the host antiviral response may be the ability of epithelial cells to generate nitric oxide (NO). NO exerts a broad range of actions, acting as a vasodilator, antimicrobial, immune regulator and neurotransmitter (Nussler and Billiar, 1993). In the last few years, attention has begun to focus on the potential role of NO in airway diseases, including asthma. Levels of NO in exhaled air from subjects with asthma are higher than those from normal subjects (Kharitonov et al., 1994; Massaro et al., 1995), but the origins and significance of exhaled NO are unclear. Indeed, it remains unclear whether increased production of NO in asthma is primarily beneficial or if it exacerbates the inflammatory state (Barnes and Liew, 1995). NO is a potent bronchodilator in human airways (Hogman et al., 1993), but NO also has been reported to have suppressive effects on the T_H1 subtype of lymphocytes (Taylor-Robinson et al., 1994), leading to possible proliferation of T_H2 cells and increased production of IL-4 and IL-5. Production of NO occurs via the actions of the enzyme NO synthase, which exists in three isoforms (Nathan and Xie, 1994), referred to as the neuronal, or type I, form (nNOS), the inducible, or type II, form (iNOS), and the constitutive, or type III, form (cNOS). It has been reported that both cNOS and iNOS can be produced from cultured respiratory epithelial cells (Asano et al., 1994; Guo et al., 1995; Shaul et al., 1994), although iNOS expression seems to predominate *in vivo*. Moreover, immunohistochemical staining of transbronchial biopsies from asthmatic patients has convincingly demonstrated increased iNOS activity compared with nonasthmatic controls (Hamid et al., 1993). Interestingly, while there is often constitutive epithelial expression of iNOS *in vivo*, expression is lost within hours of putting the cells in culture (Guo et al., 1995). In primary cells, expression can be reinduced by a variety of proinflammatory cytokines. We also have data that rhinovirus infection of primary human epithelial cells induces expression of mRNA for iNOS (Sanders and Proud, unpublished observations).

In recent years, NO has also been shown to have antiviral properties in murine cell lines and in an *in vivo* mouse model. Replication of several viruses, including vaccinia virus (Harris et al., 1995), herpes simplex-1

(Croen, 1993; Karupiah et al., 1993), vesicular stomatitis virus (Bi and Reiss, 1995), coxsackie virus (Lowenstein et al., 1996), and poliovirus (Komatsu et al., 1996) was inhibited by induction of NO synthase, or by the addition of a chemical donor of NO, S-nitroso-l-acetyl penicillamine. Indeed, in some cell types, the antiviral actions of interferon seem to be mediated via NO (Harris et al., 1995; Karupiah et al., 1993; Komatsu et al., 1996). We have now demonstrated, for the first time, that NO can inhibit both rhinovirus replication and rhinovirus induced epithelial cell cytokine production (Sanders et al., 1998). Because epithelial cells lose endogenous expression of iNOS, we have examined the effects of NO by exposing cells to a chemical donor, 3-(2-hydroxy-2-nitroso-1-propylhydrazino)-1-propanamine (NONOate) that shows pH-dependent spontaneous dissociation to release NO with a defined half life of 76 mins at pH 7.4 (Maragos et al., 1993). NONOate inhibits rhinovirus induced epithelial cell production of IL-8 and IL-6 in a dose dependent manner. Inhibition also can be observed either when NONOate is added only during the viral exposure period, or only after viral exposure, indicating that the NO is not directly killing the virus or preventing binding and internalization. As noted above, IL-8 and IL-6 are induced rapidly after rhinovirus infection of epithelial cells.

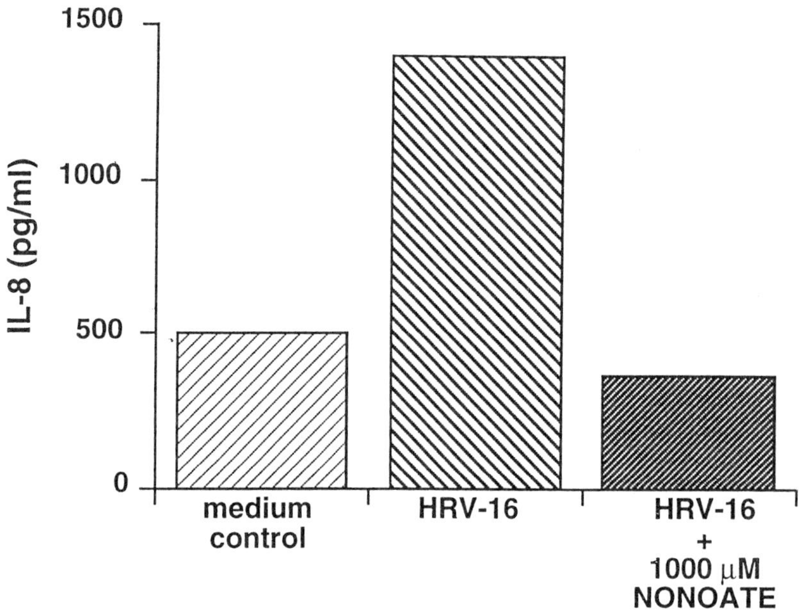

Figure 4: A nitric oxide donor (NONOate) prevents rhinovirus induced production of IL-8 from primary human epithelial cells during a 4 hour incubation period.

NONOate did not prevent the rapid viral induction of mRNA for these cytokines but inhibited production of protein products, suggesting that the effects of NO are primarily at the post-transcriptional level. The NO mediated inhibition of virally-induced IL-8 production occurred not only in immortalized epithelial cell lines, but also in primary epithelial cells (Figure 4). The effects of NONOate are due to NO because NONOate that had been depleted of its NO content was without effect. Moreover, cytokine production rebounds, as the compound gradually loses its NO content, indicating that NO must be present to continue suppressing IL-8 and IL-6 production (Sanders et al., 1998). Our more recent data also indicate that NO inhibits rhinovirus induced production of the eosinophil chemoattractant and activating cytokines, RANTES and GM-CSF. We have shown, using [^{35}S]-methionine labeling of infected cells, that the ability of NO to inhibit the production of these cytokines is a selective process and is not a result of nonspecific inhibition of protein synthesis. NO also reduced viral titers in conditioned medium recovered from above cells 24 hours after rhinovirus infection. The effects of NONOate in this regard are dose-dependent. Again, as the NONOate loses its capacity to release NO, viral replication resumes so that titers recovered during a second 24 hour collection without re-addition of NONOate are no longer different from those obtained from cells exposed to virus alone (Sanders et al., 1998).

Further study is required to delineate the mechanisms by which NO inhibits viral replication and rhinovirus induced cytokine production. Although the mechanisms by which NO inhibits rhinovirus replication are unknown, it is possible that NO functions by enhancing selected host antiviral pathways, such as activating the antiviral double-stranded RNA (DsRNA)-dependent protein kinase (PKR) (de Haro et al., 1996), or enhancing the activity of RNase L, an antiviral pathway that has been reported to be particularly active against picornaviruses (Samuel, 1991). Both of these pathways are stimulated by interferon (Lebleu et al., 1976; Samuel, 1991) and, as noted above, NO has been reported to mediate the actions of interferon in some cell types. Alternatively, NO may exert antiviral effects by blocking the actions of important viral proteins. The rhinovirus genome, for example, encodes two proteases, named 2A and 3C, which are essential in the processing of the viral polyprotein translation product and in releasing the viral RNA polymerase 3D (Rueckert, 1996). Protease 2A also helps to shut off cap-dependent host protein translation by cleaving the host eukaryotic initiation factor-4G (Kräusslich et al., 1987; Lloyd et al., 1988). At the same time, protease 2A enhances rhinovirus internal ribosome entry segment-mediated, cap-independent translation (Hambidge and Sarnow, 1992). Protease 3C cleaves the TATA-binding protein subunit of transcription factor IID, thereby also inhibiting host transcriptional processes (Clark et al., 1993). Both proteases 2A and 3C are cysteine proteases. Since NO is extremely reactive with thiol residues

(Clancy and Abramson, 1995; Stamler et al., 1992), it is possible that NO may exert its antiviral effects by inhibiting these enzymes.

The ability of NO to inhibit production of rhinovirus induced epithelial cytokines that are generated late in the viral replication cycle may, obviously, be secondary to inhibition of viral replication. Other independent pathways, however, must also be important because induction of IL-8 and IL-6 occurs early after infection and does not seem to relate to viral replication, yet induction of these cytokines is inhibited by NO (Sanders et al., 1998). Moreover, in some cells NO can inhibit cytokine production that is not induced by viral infection (Thomassen et al., 1997). The mechanisms involved in cytokine inhibition are unknown, but there is precedent for NO to affect mRNA stability (Filippov et al., 1997). Alternatively, NO may affect cytokine production, and possibly viral replication (Karupiah and Harris, 1995), via its ability to inhibit iron- and sulfur-dependent mitochondrial enzymes (e.g. aconitase and Complexes I and II of the mitochondrial electron transport chain) that are involved in energy metabolism (Nussler and Billiar, 1993; Stadler et al., 1991; Stamler et al., 1992).

VII. SUMMARY

There is now a growing body of evidence relating upper respiratory viral infections to asthma exacerbations, but the mechanisms underlying this relationship remain to be elucidated. In this chapter, we have reviewed the data supporting the hypothesis that viral induction of proinflammatory cytokine production from infected epithelial cells may play an important role in inducing asthma attacks. It is clear that cytokine production is induced from infected epithelial cells *in vitro*, and that many of the same cytokines are detected in airway secretions during viral infections *in vivo*. There is also strong evidence to support the idea that eosinophils from asthmatic and symptomatic allergic subjects may be primed to respond to the eosinophil chemoattractant cytokines, RANTES and eotaxin, that are produced from infected epithelial cells. Recently, it has been shown that ICAM-1 expression is enhanced on epithelial cells of asthmatic subjects compared to normals or to patients with chronic bronchitis, and that ICAM-1 expression correlates with asthma severity (Vignola et al., 1993). Since increasing expression of ICAM-1 on epithelial cells has been shown to enhance susceptibility of cells to infection with major group rhinoviruses (Subauste et al., 1995), asthmatic subjects may also be at greater risk than nonatopics to develop lower airway infections upon viral exposure. Increased ICAM-1 expression, together with the relative priming state of eosinophils may be important variables in the interdependence between atopic status and viral induced asthma exacerbations.

It is also apparent, however, that viral infection does not always lead to an asthma attack in asthmatic subjects. This could well relate to a complex balance between several features, including viral load and the ability of the virus to overcome the host antiviral response, such as the production of nitric oxide by epithelial cells, in order to spread to lower airway epithelial cells, as well as to the relative "primed" state of the host's circulating eosinophils. If the viral load is sufficient to induce a pronounced generation of eosinophil chemoattractant cytokines from the epithelium and to cause increased bronchial eosinophilia, then some level of enhanced airway hyperresponsiveness or reduced pulmonary function may result.

The concepts outlined in this chapter must still be validated using appropriate *in vivo* human models. In the next few years, however, it will be possible to perform experiments designed to test key elements of these hypotheses. These studies should generate new insights into the pathogenesis of viral exacerbations of asthma and may provide the basis for improved therapeutic interventions.

REFERENCES

Alam, R., York, J., Boyars, M., Stafford, S., Grant, J.A., Lee, J., Forsythe, P., Sim, T., and Ida, N. Increased MCP-1, RANTES, and MIP-1α in bronchoalveolar lavage fluid of allergic asthmatic patients. *Am. J. Respir. Crit. Care Med.* 153, 1398, 1996.

Angelini, B., Van Deusen, M. A., Doyle, W. J., Seroky, J., Cohen, S., and Skoner, D. P., Lower airway responses to rhinovirus-Hanks in healthy subjects with and without allergy. *J. Allergy Clin. Immunol.* 99, 618, 1997.

Aquilina, A. T., Hall, W. J., Douglas, R. G., and Utell, M. J., Airway reactivity in subjects with viral upper respiratory tract infections: the effects of exercise and cold air. *Am. Rev. Respir. Dis.* 122, 3, 1980.

Arnold, R., Humbert, B., Werchau, H., Gallati, H., and König, W., Interleukin-8, interleukin-6, and soluble tumour necrosis factor receptor type I release from a human pulmonary epithelial cell line (A549) exposed to respiratory syncytial virus. *Immunology* 82, 126, 1994.

Arruda, E., Boyle, T. R., Winther, B., Pevear, D. C., Gwaltney, J. M., Jr., and Hayden, F. G,. Localization of human rhinovirus replication in the upper respiratory tract by in situ hybridization. *J. Infect. Dis.* 171, 1329, 1995.

Asano, K., Chee, C. B. E., Gaston, B., Lilly, C. M., Gerard, C., Drazen, J. M., and Stamler, J. S., Constitutive and inducible nitric oxide synthase gene expression, regulation, and activity in human lung epithelial cells. *Proc. Natl. Acad. Sci. USA* 91, 10089, 1994.

Bardin, P. G., Johnston, S. L., Sanderson, G., Robinson, S., Pickett, M. A., Fraenkel, D. J., and Holgate, S. T., Detection of rhinovirus infection of the nasal mucosa by oligonucleotide in situ hybridization. *Am. J. Respir. Cell Mol. Biol.* 10, 207, 1994.

Barnes, P. J., and Liew, F. Y., Nitric oxide and asthmatic inflammation. *Immunol. Today* 16, 128, 1995.

Beasley, R., Coleman, E. D., Hermon, Y., Holst, P. E., O'Donnell, T. V., and Tobias, M., Viral respiratory tract infection and exacerbations of asthma in adult patients. *Thorax* 43, 679, 1988.

Becker, S., Koren, H. S., and Henke, D. C., Interleukin-8 expression in normal nasal epithelium and its modulation by infection with respiratory syncytial virus and cytokines tumor necrosis factor, interleukin-1 and interleukin-6. *Am. J. Respir. Cell. Mol. Biol.* 8, 20, 1993.

Bi, Z., and Reiss, C. S., Inhibition of vesicular stomatitis virus infection by nitric oxide. *J. Virol.* 69, 2208, 1995.

Bochner, B. S., and Schleimer, R. P., The role of adhesion molecules in human eosinophil and basophil recruitment. *J. Allergy Clin. Immunol.* 94, 427, 1994.

Borish, L., and Rosenwasser, L. J., Update on cytokines. *J. Allergy Clin. Immunol.* 97, 719, 1996.

Bousquet, J. B., Chanez, P., Lacoste, J. Y., Barneon, G., Ghavanian, N., Enander, I., Venge, P., Ahlstedt, S., Lafontaine, J. S., Godard, P., and Michel, F. B., Eosinophilic inflammation in asthma. *N. Engl. J. Med.* 323, 1033, 1990.

Busse, W. W., and Gern, J. E., Viruses in asthma. *J. Allergy Clin. Immunol.* 100, 147, 1997.

Calhoun, W. J., Dick, E. C., Schwartz, L. B., and Busse, W. W., A common cold virus, rhinovirus 16, potentiates airway inflammation after segmental antigen bronchoprovocation in allergic subjects. *J. Clin. Invest.* 94, 2200, 1994.

Calhoun, W. J., Swenson, C. A., Dick, E. C., Schwartz, L. B., Vrtis, R. F., and Busse, W. W., Experimental rhinovirus 16 infection potentiates histamine release after antigen provocation in allergic subjects. *Am. Rev. Respir. Dis.* 144, 1267, 1991.

Carlsen, K. H., Orstavik, I., Leegard, J., and Hoeg, H., Respiratory virus infections and aeroallergens and acute bronchial asthma. *Arch. Dis. Child.* 59, 310, 1984.

Cheung, D., Dick, E. C., Timmers, M. C., de Klerk, E. P. A., Spaan, W. J. M., and Sterk, P. J., Rhinovirus inhalation causes long-lasting excessive airway narrowing in response to methacholine in asthmatic subjects in vivo. *Am. J. Respir. Crit. Care Med.* 152, 1490, 1995.

Choi, A. M. K., and Jacoby, D. B., Influenza virus A infection induces interleukin-8 gene expression in human airway epithelial cells. *FEBS Lett.* 309, 327, 1992.

Clancy, R. M., and Abramson, S. B., Nitric oxide: a novel mediator of inflammation. *Proc. Soc. Exp. Biol. Med.* 210, 93, 1995.

Clark, M. E., Lieberman, P. M., Berk, A. J., and Dasgupta, A., Direct cleavage of human TATA-binding protein by poliovirus protease 3C in vivo and in vitro. *Mol. Cell. Biol.* 13, 1232, 1993.

Couch, R. B,. Rhinoviruses. In *Fields Virology*, Fields, B.N., Knipe, D. M., Howley, P. M., Eds., Lippincott-Raven, Philadelphia, p.713, 1996.

Croen, K. D., Evidence for an antiviral effect of nitric oxide. Inhibition of herpes simplex virus type 1 replication. *J. Clin. Invest.* 91, 2446, 1993.

de Haro, C., Méndez, R., and Santoyo, J., The eIF-2α kinases and the control of protein synthesis. *FASEB J.* 10, 1378, 1996.

DiCosmo, B. F., Geba, G. P., Picarella, D., Elias, J. A., Rankin, J. A., Stripp, B. R., Whitsett, J. A., and Flavell, R. A., Airway epithelial cell expression of interleukin-6 in transgenic mice. Uncoupling of airway inflammation and bronchial hyperreactivitity. *J. Clin. Invest.* 94, 2028, 1994.

Djukanovic, R., Roche, W., Wilson, J. W., Beasley, C. R. W., Twentyman, O. P., Howarth, P. H., and Holgate, S. T., Mucosal inflammation in asthma. *Am. Rev. Resp. Dis.* 142, 434, 1990.

Duff, A. L., Pomeranz, E. S., Gelber, L. E., Price, G. W., Farris, H., Hayden, F. G., Platts-Mills, T. A. E., and Heymann, R. W., Risk factors in acute wheezing in infants and children: viruses, passive smoke, and IgE antibodies to inhalant allergens. *Peds.* 92, 535, 1993.

Ebisawa, M., Yamada, T., Bickel, C., Klunk, D., and Schleimer, R. P., Eosinophil transendothelial migration induced by cytokines. III. Effect of the chemokine RANTES. *J. Immunol.* 153, 2153, 1994.

Einarsson, O., Geba, G. P., Zhu, Z., Landry, M., and Elias, J. A., Interleukin-11: stimulation in vivo and in vitro by respiratory viruses and induction airways hyperresponsiveness. *J. Clin. Invest.* 97, 915, 1996.

Empey, D. W., Laitinen, L., Jacobs, L., Gold, W. M., and Nadel, J. A., Mechanisms of bronchial hyperreactivity in normal subjects after upper respiratory tract infection. *Am. Rev. Respir. Dis.* 113, 131, 1976.

Evans, R., III, Mullally, D. I., Wilson, R. W., Gergen, P. J., Rosenberg, H. M., Grauman, J. S., Chevarley, F. M., and Feinleib, M., National trends in the morbidity and mortality of asthma in the US: 1965-1984. *Chest* 91, 65S, 1987.

Fahy, J. V., Kim, K. W., Liu, J., and Boushey, H. A., Prominent neutrophilic inflammation in sputum from subjects with asthma exacerbation. *J. Allergy Clin. Immunol.* 95, 843, 1995.

Farr, B. M., Gwaltney, J. M., Jr., Hendley, J. O., Hayden, F. G., Naclerio, R. M., McBride, T., Doyle, W. J., Sorrentino, J. V., Riker, D. K., and Proud, D., A randomized controlled trial of glucocorticoid prophylaxis against experimental rhinovirus infection. *J. Infect. Dis.*

162, 1173, 1990.

Feingold, B. F., Infection in bronchial allergic disease: bronchial asthma, allergic bronchitis, asthmatic bronchitis. *Pediatr. Clin. North Am.* 6, 709, 1959.

Filippov, G., Bloch, D. B., and Bloch, K. D., Nitric oxide decreases stability of mRNAs encoding soluble guanylate cyclase subunits in rat pulmonary artery smooth muscle cells. *J. Clin. Invest.* 100, 942, 1997.

Fraenkel, D. J., Bardin, P. G., Sanderson, G., Lampe, F., Johnston, S. L., and Holgate, S. T., Lower airways inflammation during rhinovirus colds in normal and asthmatic subjects. *Am. J. Respir. Crit. Care Med.* 151, 879, 1995.

Garcia-Zepeda, E. A., Rothenberg, M. E., Ownbey, R. T., Celestin, J., Leder, P., and Luster, A. D., Human eotaxin is a specific chemoattractant for eosinophil cells and provides a new mechanism to explain tissue eosinophilia. *Nature Med.* 2, 449, 1996.

Gern, J. E., Calhoun, W., Swenson, C., Shen, G., and Busse, W. W., Rhinovirus infection preferentially increases lower airway responsiveness in allergic subjects. *Am. J. Respir. Crit. Care Med.* 155, 1872, 1997.

Gern, J. E., Galagan, D. M., Jarjour, N.N., Dick, E. C., and Busse, W. W., Detection of rhinovirus RNA in lower airway cells during experimentally induced infection. *Am. J. Respir. Crit. Care Med.* 155, 1159, 1997.

Gleich, G. J., The eosinophil and bronchial asthma: Current understanding. *J. Allergy Clin. Immunol.* 85, 422, 1990.

Greve, J. M., Davis, G., Meyer, A. M., Fork, C. P., Yost, S. C., Marlor, C. W., Kamarck, M. E., and McClelland, A., The major human rhinovirus receptor is ICAM-1. *Cell* 56, 839, 1989.

Grünberg, K., Timmers, M. C., Smits, H. H., de Klerk, E. P. A., Dick, E. C., Spaan, W. J. M., Hiemstra, P. S., and Sterk, P. J., Effects of experimental rhinovirus 16 colds on airway hyperresponsiveness to histamine and interleukin-8 in nasal lavage in asthmatic subjects in vivo. *Clin. Exp. Allergy* 27, 36, 1996.

Guo, F. H., De Raeve, H. R., Rice, T. W., Stuehr, D. J., Thunnissen, F. B. J. M., and Erzurum, S. C., Continuous nitric oxide synthesis by inducible nitric oxide synthase in normal human airway epithelium in vivo. *Proc. Natl. Acad. Sci. USA* 92, 7809, 1995.

Gwaltney, J. M., Hendley, J. O., Simon, G., and Jordan, W. S., Rhinovirus infections in an industrial population. I. The occurrence of illness. *N. Engl. J. Med.* 275, 1261, 1966.

Hakansson, L., Carlson, M., Stalenheim, G., and Venge, P., Migratory responses of eosinophil and neutrophil granulocytes from asthmatic patients. *J. Allergy Clin. Immunol.* 85, 743, 1990.

Hakansson, L., Heinrich, C., Rak, S., and Venge, P., Priming of eosinophil adhesion in patients with birch pollen allergy during pollen season: Effect of immunotherapy. *J. Allergy Clin. Im.* 99, 551, 1997.

Halfon, M., and Newacheck, P. W., Trends in the hospitalization for acute childhood asthma, 1970-1984. *Am. J. Public Health* 76, 1308, 1986.

Halperin, S. A., Eggleston, P. A., Beasley, P., Suratt, P. M., Hendley, J. O., Gröschel, D. H. M., and Gwaltney, J. M., Jr., Exacerbations of asthma in adults during experimental rhinovirus infection. *Am. Rev. Respir.Dis.* 132, 976, 1985.

Halperin, S. A., Eggleston, P. A., Hendley, J. O., Suratt, P. M., Gröschel, D. H. M., and Gwaltney, J. M., Jr., Pathogenesis of lower respiratory tract symptoms in experimental rhinovirus infection. *Am. Rev. Respir. Dis.* 128, 806, 1983.

Hambidge, S., and Sarnow, P., Translational enhancement of the poliovirus 5' non-coding region mediated by virus-encoded polypeptide 2A. *Proc. Natl. Acad. Sci. USA* 89, 10272, 1992.

Hamid, Q., Springall, D. R., Riveros-Moreno, V., Chanez, P., Howarth, P., Redington, A., Bousquet, J., Godard, P., Holgate, S. T., and Polak, J. M., Induction of nitric oxide synthase in asthma. *Lancet* 342, 1510, 1993.

Harris, N., Buller, R. M. L., and Karupiah, G., Gamma interferon-induced, nitric oxide-mediated inhibition of vaccinia virus replication. *J. Virol.* 69, 910, 1995.

Haussiau, F. A., Caulie, P. G., and Van Snick, J., Distinct roles of IL-1 and IL-6 in human T cell activation. *J. Immunol.* 143, 2520, 1989.

Hofer, F., Gruenberger, M., Kowalski, H., Machat, H., Huettinger, M., Kuechler, E., and Blaas, D. Members of the low density lipoprotein receptor family mediate cell entry of a minor-group common cold virus. *Proc. Natl. Acad. Sci. USA* 91, 1839-1842, 1994.

Hogman, M., Frostell, C. G., Hedenstrom, H., and Hedenstierna, G., Inhalation of nitric oxide modulates adult human bronchial tone. *Am. Rev. Respir. Dis.* 148, 1474, 1993.

Holsti, M. A., and Raulet, D. H., IL-6 and IL-1 synergize to stimulate IL-2 production and proliferation of peripheral T cells. *J. Immunol.* 143, 2514, 1989.

Horn, M. E. C., Brain, E. A., Gregg, I., Inglis, J. M., Yealland, S. J., and Taylor, P., Respiratory viral infection and wheezy bronchitis in childhood. *Thorax* 34, 23, 1979.

Hudgel, D. W., Langston, L. J., Selner, J. C., and McIntosh, K., Viral and bacterial infections in adults with chronic asthma. *Am. Rev. Respir. Dis.* 120, 393, 1979.

Jenkins, C. R., and Breslin, A. B. X., Upper respiratory tract infections and airway reactivity in normal and asthmatic subjects. *Am. Rev. Respir. Dis.* 130, 879, 1984.

Jennings, L. C., Barns, G., and Dawson, K. P., The association of viruses

with acute asthma. *NZ Med. J.* 100, 488, 1987.
Johnston, S. L., Pattemore, P. K., Sanderson, G., Smith, S., Campbell, M. J., Josephs, L. K., Cunningham, A., Robinson, B. S., Myint, S. H., Ward, M. E., Tyrrell, D. A. J., and Holgate, S. T., The relationship between upper respiratory infections and hospital admissions for asthma: a time-trend analysis. *Am. J. Respir. Crit. Care Med.* 154, 654, 1996.
Johnston, S. L., Pattemore, P. K., Sanderson, G., Smith, S., Lampe, F., Josephs, L., Sympington, P., O'Toole, S., Myint, S. H., Tyrrell, D. A., and Holgate, S. T., Community study of role of viral infections in exacerbations of asthma in 9-11 year old children. *Br. Med. J.* 310, 1225, 1995.
Karupiah, G., and Harris, N., Inhibition of viral replication by nitric oxide and its reversal by ferrous sulfate and tricarboxylic acid cycle metabolites. *J. Exp. Med.* 181, 2171, 1995.
Karupiah, G., Xie, Q.-W., Buller, R. M. L., Nathan, C., Duarte, C., and MacMicking, J. D,. Inhibition of viral replication by interferon-γ-induced nitric oxide synthase. *Science* 261, 1445, 1993.
Kharitonov, S. A., Yates, D., Robbins, R. A., Logan-Sinclair, R., Shinebourne, E., and Barnes, P. J., Increased nitric oxide in exhaled air of asthmatic patients. *Lancet* 343, 133, 1994.
Komatsu, T., Bi, Z., and Reiss, C. S., Interferon-γ induced type I nitric oxide synthase activity inhibits viral replication in neurons. *J. Neuroimmunol.* 68, 101, 1996.
Kräusslich, H. G., Nicklin, M. J., Toyoda, H., Etchison, D., and Wimmer, E., Poliovirus protease 2A induces cleavage of eularyotic initiation factor 4F polypeptide p220. *J. Virol.* 61, 1478, 1987.
Laitinen, L. A., Laitinen, A., Heino, M., and Haahtela, T., Eosinophilic airway inflammation during exacerbation of asthma and its treatment with inhaled corticosteroid. *Am. Rev. Respir. Dis.* 143, 423, 1991.
Larsen, C. G., Anderson, A. O., Appella, E., Oppenheim, J. J., and Matsushima, K., The neutrophil activating protein (NAP-1) is also chemotactic for T lymphocytes. *Science* 243, 1464, 1989.
Larson, H. E., Reed, S. E., and Tyrrell, D. A. J., Isolation of rhinoviruses and coronaviruses from 38 colds in adults. *J. Med. Virol.* 5, 221, 1980.
Lebleu, B., Sen, G. C., Shaila, S., Cabrer, B., and Lengyel, P., Interferon, double-stranded RNA, and protein phosphorylation. *Proc. Natl. Acad. Sci. USA* 73, 3107, 1976.
Lemanske, R. F., Dick, E. C., Swenson, C. A., Vrtis, R. F., and Busse, W. W., Rhinovirus upper respiratory infection increases airway hyperreactivity and late asthmatic reactions. *J. Clin. Invest.* 83, 1, 1989.
Leonard, E. J., and Yoshimura, T., Neutrophil attractant/activation protein-1 (NAP-1 [interleukin-8]). *Am. J. Respir. Cell Mol. Biol.* 2, 479, 1990.
Levandowski, R. A., Ou, D. W., and Jackson, G. G., Acute-phase decrease

of T lymphocyte subsets in rhinovirus infection. *J. Infect. Dis.* 153, 743, 1986.

Levandowski, R. A., Weaver, C. W., and Jackson, G. G., Nasal secretion leukocyte populations determined by flow cytometry during acute rhinovirus infection. *J. Med. Virol.* 25, 423, 1988.

Lloyd, R. E., Jense, H. G., and Ehrenfeld, E., Relationship of p220 cleavages during picornavirus infection to 2A protease sequences. *J. Virol.* 62, 4216, 1988.

Lopez, A. F., Williamson, D. J., Gamble, J. R., Begley, C. G., Harlan, J. M., Klebanoff, S. J., Waltersdorph, A., Wong, G., Clark, S. C., and Vadas, M. A., Recombinant human granulocyte-macrophage colony-stimulating factor stimulates in vitro mature human neutrophil and eosinophil function, surface receptor expression, and survival. *J. Clin. Invest.* 78, 1220, 1986.

Lowenstein, C. J., Hill, S. L., Lafond-Walker, A., Wu, J., Allen, G., Landavere, M., Rose, N. R., and Herskowitz, A., Nitric oxide inhibits viral replication in murine myocarditis. *J. Clin. Invest.* 97, 1837, 1996.

Maragos, C. M., Wang, J. M., Hrabie, J. M., Oppenheim, J. J., and Keefer, L. K., Nitric oxide/nucleophile complexes inhibit the in vitro proliferation of A375 melanoma cells via nitric oxide release. *Cancer Res.* 53, 564, 1993.

Martinez, F. D., Wright, A. L., Taussig, L. M., Holberg, C. J., Halonen, M., Morgan, W. J., and Group Health Medical Associates, Asthma and wheezing in the first six years of life. *N. Engl. J. Med.* 332, 133, 1995.

Massaro, A. F., Gaston, B., Kita, D., Fanta, C., Stamler, J. S., and Drazen, J. M., Expired nitric oxide levels during treatment of acute asthma. *Am. J. Respir. Crit. Care Med.* 152, 800, 1995.

Matsukura, S., Kokubo, F., Noda, H., Tokunaga, H., and Adachi, M., Expression of IL-6, IL-8, and RANTES on human bronchial epithelial cells, NCI-H292, induced by influenza virus A. *J. Allergy Clin. Immunol.* 98, 1080, 1996.

McIntosh, K., Ellis, E. F., Hoffman, L. S., Lybass, T. G., Eller, J. J., and Fulginiti, V. A., The association of viral and bacterial respiratory infections with exacerbations of wheezing in your asthmatic children. *J. Pediatr.* 82, 578, 1973.

Mitchell, I., Inglis, H., and Simpson, H., Viral infections in wheezy bronchitis and asthma in children. *Arch. Dis. Child.* 51, 707, 1976.

Naclerio, R. M., Proud, D., Lichtenstein, L. M., Kagey-Sobotka, A., Hendley, J. O., Sorrentino, J., and Gwaltney, J. M., Kinins are generated during experimental rhinovirus colds. *J. Infect. Dis.* 157, 133, 1988.

Nathan, C., and Xie, Q. W., Nitric oxide synthase: roles, tolls and controls. *Cell* 78, 915, 1994.

Nicholson, K. G., Kent, J., and Ireland, D. C., Respiratory viruses and exacerbations of asthma in adults. *Br. Med. J.* 307, 982, 1993.

Noah, T. L., and Becker, S., Respiratory syncytial virus-induced cytokine production by a human bronchial epithelial cell line. *Am. J. Physiol.* 265, L472, 1993.

Noah, T. L., Henderson, F. W., Wortman, I. A., Devlin, R. B., Handry, J., Koren, H. S., and Becker, S., Nasal cytokine production in viral acute upper respiratory infection of childhood. *J. Infect. Dis.* 171, 584, 1995.

Nussler, A. K., and Billiar, T. R., Inflammation, immunoregulation, and inducible nitric oxide synthase. *J. Leukocyte Biol.* 54, 171, 1993.

Owen, W. F., Jr., Rothenberg, M. E., Silberstein, D. S., Gasson, J. C., Stevens, R. L., and Austen, K. F., Regulation of human eosinophil viability, density, and function by granulocyte/macrophage colony-stimulating factor in the presence of 3T3 fibroblasts. *J. Exp. Med.* 166, 129, 1987.

Pattemore, P. K., Johnston, S. L., and Bardin, P. G., Viruses as precipitants of asthma symptoms. I. Epidemiology. *Clin. Exp. Allergy* 22, 325, 1992.

Podosin, R. L., and Felton, W. L., The clinical picture of a Far-East influenza occurring at the 4th National Boy Scout Jamboree. *N. Engl. J. Med.* 238, 778, 1958.

Polito, A. J., and Proud, D., Epithelial cells as inflammatory cells. In *Asthma*, Barnes, P. J., Grunstein, M. M., Leff, A. R., and Woolcock, A. J., Eds,. Lippincott-Raven Publishers, Philadelphia, p.491, 1997.

Ponath, P. D., Qin, S., Ringler, D. J., Clark-Lewis, I., Wang, J., Kassam, N., Smith, H., Shi, X., Gonzalo, J.-A., Newman, W., Gutierrez-Ramos, J.-C., and Mackay, C. R., Cloning of the human eosinophil chemoattractant, eotaxin. *J. Clin. Invest.* 97, 604, 1996.

Potter, P. C., Weinberg, E., and Shore, S. C. L., Acute severe asthma: a prospective study of the precipitating factors in 40 children. *S. Afr. Med. J.* 66, 397, 1984.

Proud, D., Gwaltney, J. M., Jr., Hendley, J. O., Dinarello, C. A., Gillis, S., and Schleimer, R. P., Increased levels of interleukin-1 are detected in nasal secretions of volunteers during experimental rhinovirus colds. *J. Infect. Dis.* 169, 1007, 1994.

Proud, D., Naclerio, R. M., Gwaltney, J. M., and Hendley, J. O., Kinins are generated in nasal secretions during natural rhinovirus colds. *J. Infect. Dis.* 161, 120, 1990.

Rebhan, A. W., An outbreak of Asian influenza in a girls' camp. *Can. Med. Assoc. J.* 77, 797, 1957.

Rot, A., Krieger, M., Brunner, T., Bischoff, S. C., Schall, T. J., and Dahinden, C. A., RANTES and macrophage inflammatory protein 1a induce the migration and activation of normal human eosinophil granulocytes. *J. Exp. Med.* 176, 1489, 1992.

Rueckert, R. R., Picornaviridae: the viruses and their replication. In *Fields Virology*, Fields, B.N., Knipe, D. M., and Howley, P. M., Eds., Lippincott-Raven, Philadelphia, p.609, 1996.
Saito, T., Deskin, R. W., Casola, A., Häeberle, H., Olszewska, B., Ernst, P. B., Alam, R., Ogra, P. L., and Garofalo, R., Respiratory scyncytial virus induces selective production of the chemokine RANTES by upper airway epithelial cells. *J. Infect. Dis.* 175, 497, 1997.
Samuel, C. E., Antiviral actions of interferon. Interferon-regulated cellular proteins and their surprisingly selective antiviral activities. *Virology* 183, 1, 1991.
Sanders, S. P., Siekierski, E. S., Porter, J. D., Richards, S. M., and Proud, D., Nitric oxide inhibits rhinovirus-induced cytokine production and viral replication in a human respiratory epithelial cell line. *J. Virol.*, 72, 934, 1998.
Schall, T. J., Bacon, K., Toy, K. J., and Goeddel, D. V., Selective attraction of monocytes and T lymphocytes of the memory phenotype by cytokine RANTES. *Nature* 347, 669, 1990.
Shaul, P. W., North, A. J., Wu, L. C., Wells, L. B., Brannon, T. S., Lau, K. S., Michel, T., Margraf, L. R., and Star, R. A., Endothelial nitric oxide synthase is expressed in cultured human bronchiolar epithelium. *J. Clin. Invest.* 94, 2231, 1994.
Skoner, D. P., Doyle, W. J., Tanner, E. P., Kiss, J., and Fireman, P., Effects of rhinovirus 39 (RV-39) infection on immune and inflammatory parameters in allergic and non-allergic subjects. *Clin. Exp. Allergy* 25, 561, 1995.
Sly, R. M., Mortality from asthma, 1979-1984. *J. Allergy Clin. Immunol.* 82, 705, 1988.
Smith, D. H., Malone, D. C., Lawson, K. A., Okamoto, L. J., Battista, C., and Saunders, W. B., A national estimate of the economic costs of asthma. *Am. J. Respir. Crit. Care Med.* 156, 787, 1997.
Stadler, J., Curran, R. D., Ochoa, J. B., Harbreacht, B. G., Hoffman, R. A., Simmons, R. L., and Billiar, T. R., Effect of endogenous nitric oxide on mitochondrial respiration of rat hepatocytes in vitro and in vivo. *Arch. Surg.* 126, 186, 1991.
Stamler, J. S., Singel, D. J., and Loscalzo, J., Biochemistry of nitric oxide and its redox-activated forms. *Science* 258, 1898, 1992.
Staunton, D. E., Merluzzi, V. J., Rothlein, R., Barton, R., Marlin, S. D., and Springer, T. A., A cell adhesion molecule, ICAM-1, is the major surface receptor for rhinoviruses. *Cell* 56, 849, 1989.
Storr, J., and Lenney, W., School holidays and admissions with asthma. *Arch. Dis. Child.* 64, 103, 1989.
Subauste, M. C., Jacoby, D. B., Richards, S. M., and Proud, D., Infection of a human respiratory epithelial cell line with rhinovirus. *J. Clin. Invest.* 96, 549, 1995.

Tang, W., Geba, G. P., Zheng, T., Ray, P., Homr, R. J., Kuhn, C., III., Flavell, R. A., and Elias, J. A., Targeted expression of IL-11 in the murine airway causes lymphocytic inflammation, bronchial remodeling and airways obstruction. *J. Clin. Invest.* 98, 2845, 1996.

Taylor-Robinson, A. W., Liew, F. Y., and Severin, A., Regulation of the immune response by nitric oxide differentially produced by T helper type 1 and T helper type 2 cells. *Eur. J. Immunol.* 24, 980, 1994.

Terajima, M., Yamaya, M., Sekizawa, K., Okinaga, S., Suzuki, T., Yamada, N., Nakayama, K., Ohrui, T., Oshima, T., Numazaki, Y., and Sasaki, H., Rhinovirus infection of primary cultures of human tracheal epithelium. *Am. J. Physiol.* 273, L749, 1997.

Teran, L. M., Johnston, S. L., Schröder, J.-M., Church, M. K., and Holgate, S. T., Role of nasal interleukin-8 in neutrophil recruitment and activation in children with virus-induced asthma. *Am. J. Respir. Crit. Care Med.* 155, 1362, 1997.

Thelen, M., Peveri, P., Kernen, P., von Tscharner, V., Walz, A., and Baggiolini, M., Mechanism of neutrophil activation by NAF, a novel monocyte-derived peptide agonist. *FASEB J.* 2, 2702, 1988.

Thomassen, M. J., Buhrow, L. T., Connors, M. J., Kaneko, F. T., Erzurum, S. C., and Kavuru, M. S., Nitric oxide inhibits inflammatory cytokine production by human alveolar macrophages. *Am. J. Respir. Cell Mol. Biol.* 17, 279, 1997.

Trigg, C. J., Nicholson, K. G., Wang, J. H., Ireland, D. C., Jordan, S., Duddle, J. M., Hamilton, S., and Davies, R. J., Bronchial inflammation and the common cold: a comparison of atopic and non-atopic individuals. *Clin. Exp. Allergy* 26, 665, 1996.

Turner, R. B., Hendley, J. O., and Gwaltney, J. M., Jr., Shedding of infected epithelial cells in rhinovirus colds. *J. Infect. Dis.* 145, 849, 1982.

Venge, J., Lampinen, M., Hakansson, L., Rak, S., and Venge, P., Identification of IL-5 and RANTES as the major eosinophilic chemoattractants in the human lung. *J. Allergy Clin. Immunol.* 97, 1110, 1996.

Vignola, A. M., Campbell, A. M., Chanez, P., Bousquet, J., Paul-Lacoste, P., Michel, F.B., and Godard, P., HLA-DR and ICAM-1 expression on bronchial epithelial cells in asthma and chronic bronchitis. *Am. Rev. Respir. Dis.* 148, 689, 1993.

Wardlaw, A. J., Dunnette, S., Gleich, G. J., Collins, J. V., and Kay, A. B., Eosinophils and mast cells in bronchoalveolar lavage in subjects with mild asthma. *Am. Rev. Respir. Dis.* 137, 62, 1988.

Warringa, R. A. J., Mengelers, H. J. J., Raaijmakers, J. A. M., Bruijnzeel, P. L. B., and Koenderman, L., Upregulation of formyl-peptide and interleukin-8 induced eosinophil chemotaxis in patients with allergic asthma. *J. Allergy Clin. Immunol.* 91, 1198, 1993.

Weiss, K. B., Gergen, P. J., and Hodgson, T. A., An economic evaluation of asthma in the United States. *N. Engl. J. Med.* 326, 862, 1992.

Winther, B., Brofeldt, S., Christensen, B., and Mygind, N., Light and scanning electron microscopy of nasal biopsy material from patients with naturally acquired common colds. *Acta. Otolaryngol. (Stockh)* 97, 309, 1984a.

Winther, B., Farr, B., Turner, R. B., Hendley, J. O., Gwaltney, J. M., Jr., and Mygind, N., Histopathologic examination and enumeration of polymorphonuclear leukocytes in the nasal mucosa during experimental rhinovirus colds. *Acta. Otolaryngol. Suppl. (Stockh)* 413, 19, 1984b.

Zhu, Z., Tang, W., Gwaltney, J. M., Jr., Wu, Y., and Elias, J. A., Rhinovirus stimulation of interleukin-8 in vivo and in vitro: role of NF-kB. *Am. J. Physiol.* 273, L814, 1997.

Zhu, Z., Tang, W., Ray, A., Wu, Y., Einarsson, O., Landry, M. L., Gwaltney, J. M., Jr., and Elias, J. A., Rhinovirus stimulation of interleukin-6 in vivo and in vitro. Evidence for nuclear factor kB-dependent transcriptional activation. *J. Clin. Invest.* 97, 421, 1996.

Chapter 3

THE RISK OF DEVELOPING ASTHMA DURING CHILDHOOD

Peter W. Heymann, M.D., Alessandro de Alarcon, B.A., Juan C. Zambrano, M.D. and Gary P. Rakes, M.D.

I. INTRODUCTION

The prevalence of asthma among school-aged children has been estimated to be 7-10% (Gergen et al., 1988; Weitzman et al., 1992; Burney, 1997). However, the incidence and severity of this disorder in the United States and other Western countries has been increasing for at least 2 decades (Gergen et al., 1990). According to prospective studies, almost half of school children who develop asthma begin having problems during their pre-school years (Sporik et al., 1990; Martinez et al., 1995). The effort to diagnose asthma early in life remains an important research priority in order to initiate treatment to diminish the frequency and severity of attacks. Studies directed toward achieving this objective have examined the relationship between infantile wheezing and asthma and have focused on environmental antigens and irritants (e.g. allergens, viral respiratory tract infections, and passive smoke) which are known to initiate or aggravate airway inflammation. Unless individuals are genetically predisposed to develop asthma, it is doubtful that any of these environmental exposures will have detrimental effects on the developing lung or immune system. However, in those who are genetically susceptible, the timing of relevant antigen exposures during early life may be critical to the recruitment of genes involved in asthma pathogenesis and there is recent speculation that events influencing this process may even begin *in utero* or during the early post-natal period (Brown et al., 1997).

II. THE RELATIONSHIP BETWEEN EARLY CHILDHOOD WHEEZING AND ASTHMA

Hospitalization rates for wheezing are one of the parameters used for monitoring the changing incidence of asthma during childhood. Curiously, hospitalizations have risen most dramatically for children under 4 years of age when the diagnosis of asthma is most challenging (Gergen et al., 1990). Within this age group are infants who are admitted to the hospital for wheezing caused by the respiratory syncytial virus (RSV) (Glezen et al., 1973). However, the pathogenesis and risk factors for wheezing in infants (also called bronchiolitis) may differ substantially from those which put children at risk for exacerbations of asthma (Heymann et al., 1998). In

1-56670-373-5/99/$0.00+$.50

this regard, smaller lung volumes at birth (detected more frequently in males than females at full-term) have been shown to predispose children to wheeze with viral infections during the first year of life, but are not predictive of asthma (Martinez et al., 1988). Other newborns who can have small lung volumes and an increased risk for infantile wheezing are those who are born prematurely, full-term infants who are small for gestational age, and those whose mothers smoked during pregnancy (Tager et al., 1993).

There are other major differences which distinguish early childhood wheezing and asthma. First, RSV is isolated in up to 70% of nasal secretions from children who wheeze during the first two years of life (Duff et al., 1993). In contrast, rhinovirus is the most frequent viral pathogen associated with wheezing after the age of two (Duff et al., 1993). Second, it has been difficult to demonstrate a link between infantile wheezing and allergic risk factors which play a dominant role in the evolution of asthma. Using conventional tests to measure IgE antibody, it is evident that sensitization to common aeroallergens is diminished during the first 2 years of life. In emergency room studies of wheezing infants, the prevalence of IgE antibody to inhaled allergens in serum was very low using the radioallergosorbent (RAST) test (Heymann et al., 1995). The same observation has also been reported in perspective studies of infants born to allergic parents by RAST and skin test analyses (van Asperen et al., 1984; Rowntree et al., 1985). After the age of 2 years, however, a significant proportion of children who are treated for wheezing (approximately 50%) have allergen specific IgE antibody and/or elevated total IgE in their serum (Heymann et al., 1995). By the age of 4 years, a larger proportion of wheezing children, approaching 80%, are sensitized to allergens compared to 20-25% in the general population (Sporik et al., 1991; Heymann et al., 1995; Bousquet et al., 1986). Thus, a dichotomy is apparent between the risk factors for wheezing during infancy and wheezing which begins or continues after the age of two years (Heymann et al., 1998). Moreover, the prognosis for infantile wheezing is generally good and, by itself, does not have much value in predicting the development of asthma. This was apparent in a prospective study of children born to allergic parents. Although these children have an increased risk for developing asthma, the majority who wheezed as infants were no longer wheezing during their toddler years (Sporik et al., 1991). In another study, 60% of children who experienced wheezing during the first three years of life, irrespective of family history, were no longer wheezing by the age of 6 (Martinez et al., 1995).

III. DEVELOPMENT OF SENSITIZATION TO ALLERGENS

The well-documented relationship between allergen sensitization (IgE antibody) and asthma has stimulated investigations of immune responses to allergens which evolve during early childhood and even prenatally in the development of the fetus. The production of IgE immunoglobulin has been detected in the fetal liver and lungs as early as eleven weeks gestation (Miller et al., 1973). Prenatal synthesis of IgE antibody to parasite antigens has also been reported (Weil et al., 1983). At birth, the capacity of newborn B cells to produce IgE has been observed in cord blood mononuclear cells stimulated with recombinant IL-4 (O'Keefe et al., 1991). Moreover, the possibility that T cells may already be sensitized by food and aeroallergens *in utero* has been proposed based on studies which report the proliferation of cord blood mononuclear cells and cytokine production following stimulation with cow milk proteins (alpha-lactalbumin, beta-lactoglobulin and alpha-casein) and Der p I and Lol p I aeroallergens (Szépfalusi et al., 1997; Piccinni et al., 1993). However, the nature of the antigenic stimulus which might cause T cell sensitization prenatally (e.g. peptides processed by the mother, or antigen: IgG complexes which might gain access to the fetal circulation) is uncertain (Brown et al., 1997). It is also not clear whether these responses in cord blood have the potential to identify newborns who will become allergic or asthmatic.

Other investigators have proposed that sensitization to inhaled allergens may begin shortly after birth because of positive correlations observed between allergen exposures present during the month of birth and the development of sensitization to the same allergens later in childhood (e.g. birch, grass, and dust mite allergens) (Bjorksten et al., 1980; Smith et al., 1979; Kemp et al., 1979; Carosso et al., 1986). The possibility of sensitization during early childhood was also suggested in a study of infants from allergic families. Those who developed asthma were more likely to wheeze at an early age if they were exposed to high levels of mite allergen measured in dust samples at one year of age (Sporik et al., 1990). In the same study however, it was difficult to detect the presence of serum IgE antibody to mite allergen by RAST before two years of age.

Results from these studies raise the question as to whether skin tests or *in vitro* assays for IgE antibody are sensitive enough to monitor the development of sensitization, and support the hypothesis that this process may begin in the respiratory tract before it can be detected in the blood or on skin mast cells. It has also been proposed that a network of MHC class II expressing dendritic cells in the lungs is likely to play a central role in the initiation of immune responses to inhaled allergens and strongly influence the Th_1 vs. Th_2 selection process (Holt et al., 1997). The pathogenic effects of other environmental stimuli on the development of allergen sensitization and asthma has also been considered. The possibility that viral respiratory tract infections may influence the timing and prevalence of allergen sensitization during early childhood has been proposed (Frick, 1983). It has also been suggested that higher rates of viral infection among children raised in urban industrialized environments may actually diminish the prevalence of allergic disorders and asthma (von Mutius et al., 1994; Bjorksten, 1997). The premise of these latter studies is that recurrent exposure to viral antigens during early childhood may modify and deviate host immune response to inhaled allergens toward a Th_1 lymphocyte reaction. However, other studies suggest that allergen specific Th_2 lymphocytes may already be established *in utero* or during the newborn period prior to any exposure to respiratory tract infections (Piccinni et al., 1993; Szépfalusi et al., 1997; Warner et al., 1997). Chronic exposure to tobacco smoke, particularly to smoke from mothers, has been shown to increase the risk for more frequent and persistent exacerbations of wheezing in infants and older children (Murray et al., 1986; Young et al., 1991). During infancy, the risk for virus-induced wheezing is increased among children whose mothers smoked during pregnancy (Jager et al., 1993). Postnatally, we have observed marked elevations of cotinine, a nicotine metabolite, in saliva from young children exposed to passive smoke at home up to the age of 4 years (Chang et al., 1995). In the same study, smoke from mothers was strongly associated with high levels of cotinine during the first 2 years of life. Whether passive smoke exposure also influences the timing of allergen sensitization remains controversial (Ownby et al., 1988).

In contrast to some of the uncertainties about aeroallergen sensitization during the first few months of life, sensitization to common food allergens can be clearly demonstrated by skin test and RAST as early as 2 to 3 months of age in up to 30% of infants born to allergic parents (Rowntree et al., 1985). This response can also be demonstrated in babies who are exclusively breast fed and who are exposed to very small amounts of allergen compared to infants on an open diet (Zeiger et al., 1989). Increased gut permeability during the newborn period and differences in antigen presentation in the gastrointestinal tract have been offered as explanations, but it is clear from this observation that children can establish a Th_2 lymphocyte response to ingested allergens early in life.

The most common clinical manifestation of food hypersensitivity in young children is atopic dermatitis (allergic eczema). Among infants who develop IgE antibody to food allergens with or without atopic dermatitis, a large percentage (80-90%) will subsequently become sensitized to aeroallergens (often during their toddler years) and develop allergic respiratory tract symptoms (Buffum et al., 1966; Langeland, 1983; Guillet et al., 1992). This has been well documented for infants with atopic dermatitis who require frequent applications of topical steroids to control skin inflammation (i.e. moderate to severe eczema), who not uncommonly become sensitized to aeroallergens before a year of age (Guillet et al, 1992; Duff et al., 1992). However, it is also true that these children represent a minority of those who will eventually develop asthma. In an emergency room study of acute wheezing, only 1/3 of the children treated for asthma had evidence of serum IgE antibody to common food allergens, and only 10% had a history of eczema during infancy (Price et al., 1995). Thus, it appears that the majority of children who become asthmatic do so without developing persistent sensitization or symptoms related to food allergens.

IV. TOTAL SERUM IGE

Determinations of total IgE generally do not represent a sensitive screening tool for judging whether individual children are atopic. This is particularly true during the first two years of life when total IgE levels remain quite low, except in children with atopic dermatitis who can have impressive titers of IgE in their serum. Moreover, higher levels of IgE in cord blood have also been observed in newborns who developed atopic dermatitis during infancy (Halonen et al., 1992). However, the screening efficiency of cord blood IgE to predict wheezing in infancy, or to predict the development of allergic disease later in childhood is too low to warrant the use of this test to identify neonates who might be candidates for allergy prevention measures (Chandra et al., 1985; Croner et al., 1990; Ruiz et al., 1990; van der Ent et al., 1990). In population based studies a positive correlation between total IgE levels and asthma has been demonstrated in groups of older children and adults (Burrows et al., 1989). This has also been shown for geometric mean levels of total IgE in infants, toddlers, and school-aged children who were treated for wheezing in an emergency room (Heymann et al., 1995). In another study, group mean levels of total IgE were shown to be higher at 9 months among wheezing infants who still experienced attacks at age 6 compared to those who wheezed, but then became asymptomatic as they grew older (Martinez et al., 1995, a). In the same patients, the production of IFN-gamma was assessed in mitogen stimulated mononuclear cells. Interferon-gamma production in cells from peripheral blood at age 9 months, but not at birth in cells from cord blood, was found to be inversely related to the presence of positive skin tests to aeroallergens in these children by age 6 years and also to positive skin test reactivity in their parents (Martinez et al., 1995, b).

V. GENETIC LINKAGE BETWEEN ATOPY AND ASTHMA

The strong association between allergic diseases and asthma, and their clustering in members of the same family, has stimulated research regarding the relationship between genes which regulate the clinical expression of these disorders (Rosenwasser et al., 1997). A significant association has been reported between total serum IgE concentrations and a cluster of genes on the long arm of chromosome 5 (5q31.1) which includes candidate genes that express the production of IL-4, IL-5, and the beta-2 adrenergic receptor (Marsh et al., 1994). However, associations with other genetic loci have also been reported and it is doubtful that the genetic regulation of allergic diseases and asthma will be explained by a single, or even a few genes that will simplify efforts to screen for these disorders (Cookson et al., 1989; Shirakawa et al., 1994). Moreover, although the genotype of an individual may determine their potential for developing asthma and/or other allergic diseases, environmental stimuli are likely to strongly influence the onset and severity of symptoms. Thus, the detection of early childhood asthma will most likely require genetic screening combined with sensitive phenotypic markers central to the pathogenesis of asthma in order to achieve early diagnosis.

VI. EARLY CHILDHOOD WHEEZING AND EOSINOPHIL RESPONSES

The recruitment of eosinophils and the toxic effect of their mediators in the respiratory tract is critical to the late-phase inflammatory events of asthma. However, there is limited information about the role of eosinophil activation and degranulation in the lower respiratory tract of infants who wheeze. Among children younger than age two who were treated for wheezing in an emergency room, eosinophil counts in peripheral blood did not differ significantly from eosinophil counts in controls (Ingram et al., 1995). In addition, high total blood eosinophil counts over 500 cells/mm^3, which are observed in older children and adults during attacks of asthma, are not often seen in wheezing infants. It has also been difficult to visualize eosinophils in nasal secretions from actively wheezing infants (Duff et al., 1993). By one year of age, however, these cells were seen in increased numbers in scrapings from the nasal mucosa of children who developed allergic respiratory tract symptoms (predominantly allergic rhinitis) by the age of four years (Zeiger et al., 1993).

A more prominent role for eosinophils in the pathogenesis of infantile wheezing has been proposed based on measurements of eosinophil mediators in nasal washes. Initially, elevated levels of eosinophil cationic protein (ECP) were described in a study of infants infected with RSV and higher levels were reported in those who wheezed than in those with rhinitis (Garofalo et al., 1992). More recently, high ECP levels were measured in nasal washes from 41% of wheezing children who were under two years of age; however, increased levels were also detected in those with rhinitis and not all infants with high ECP were infected with RSV (Ingram et al., 1995; La Russa et al., 1996). The stimuli for eosinophil activation and degranulation in these virus infected infants is unknown and the predictive value of elevated nasal ECP for the development of asthma is also unclear. Although increased concentrations of ECP in serum have been reported to correlate with recurrent wheezing when infants are followed for one year after their first attack (Koller et al., 1997), these children will need to be followed longer (i.e into their school-age years) to learn whether this finding is truly predictive of asthma.

VII. WHEEZING AND VIRAL INFECTIONS DURING INFANCY

During the first two years of life, wheezing is experienced by 20-30% of children when they acquire a viral respiratory tract infection. Many of these wheezing episodes (approximately 70%) are caused by RSV, and most hospitalizations for RSV induced wheezing occur in infants who are less than six months old (Duff et al., 1993; LaVia et al., 1992) Although the majority of these children will not develop asthma, it is estimated that up to a third of those who wheeze during the first two years of life will continue to wheeze as they grow older (Morgan et al., 1992). As a result, there is considerable interest as to whether RSV can stimulate different (e.g. T cells of the Th_2 phenotype) immune responses in those who are genetically predisposed to develop asthma. If so, the detection of this altered immune response may provide insight into the pathogenesis and diagnosis of early childhood asthma.

The capacity of RSV to induce changes characteristic of asthma was initially proposed based on the detection of IgE antibodies specific for viral antigen in naso-pharyngeal secretions from wheezing infants (Welliver et al., 1981). In that study the titers of IgE to RSV were reported to correlate with the severity of wheezing as judged by measurements of arterial pO_2 and also with nasal histamine concentrations. However, this antibody response did not prove to be predictive of asthma when these children were re-examined at the age of 7 years (Welliver et al., 1993). The detection of IgE antibody to the immunodominant RSV capsular antigens (F and G proteins) has also been reported in nasal washes by the same authors (Welliver et al., 1989.) We have also detected IgE antibody to both the RSV F and G antigens in nasal washes from infants (de Alarcon et al., 1997). However, the titers which were measured were very low compared to titers of IgG and IgA antibody to these proteins and the presence of IgE antibody to F and G was not limited to wheezing patients and did not correlate with increased levels of nasal eosinophil cationic protein as a marker of allergic inflammation. Based on these results, the clinical significance of this response is questionable.

More serious lower respiratory tract disease has also been reported in infants who acquired RSV infection after they were immunized with a formalin-inactivated RSV vaccine in the 1970's. Curiously, over half of the vaccine recipients developed peripheral blood eosinophilia when they became infected, and lung segments from some of those who died also showed pulmonary eosinophil infiltration (Chin et al., 1969; Kim et al.,

1969). Additional evidence for the capacity of RSV antigen to stimulate an eosinophil response in the lungs has been observed in animal experiments. When mice were immunized with the RSV-G (attachment) protein they developed pulmonary eosinophilia after intra-nasal inoculation and infection with RSV (Openshaw et al., 1992; Srikiatkhachorn et al., 1997). Lymphocytes from these mice also released IL-5 and reduced levels of IFN-gamma when stimulated with RSV-G protein in culture. In contrast, mice immunized with the RSV-F (fusion) protein preferentially developed a Th_1 lymphocyte response characterized by increased production of IFN-gamma in culture and little evidence for pulmonary eosinophilia following RSV infection.

These observations have stimulated interest in the relationship between RSV infection and the development of asthma. One should remember, however, that most episodes of wheezing with RSV occur during the first six months of life when many children are likely to be experiencing their first infection with this virus. This differs from the pre-sensitized mice and humans discussed above. Moreover, infants who are born full term are also likely to have neutralizing IgG antibody to RSV acquired from their mothers and it is possible that this antibody modulates the initial sensitization process. The possibility that RSV may stimulate an eosinophil response during subsequent infections remains an open question. However, as children grow older RSV infections are more likely to cause symptoms in the upper respiratory tract. Thus, although repeated infections with RSV may initiate a Th_2 lymphocyte response and eosinophilia in early childhood, it is not clear that this process plays an important role in the development of asthma. It has also been proposed that vigorous stimulation of CD8 + lymphocytes by repeated RSV infections may downregulate the Th_2 response and facilitate a shift toward a more protective Th_1 response (Srikiatkhachorn et al., 1997). Consistent with this, rhinovirus, rather than RSV, becomes the main viral pathogen associated with acute wheezing from the age of 2 on, when recurrent wheezing is a stronger predictor of asthma (Duff et al., 1993; Heymann et al., 1998).

VIII. CONCLUSION

Many children who develop asthma begin having symptoms during their pre-school years. As a result, there is good reason to promote research devoted to genetic and environmental factors which are central to the pathogenesis of this chronic childhood disorder. Results from studies of the developing immune system indicate that CD4 + T helper cells may initially be biased toward the production of Th_2 cytokines (Wilson et al., 1986; Lin et al., 1993). Moreover, it is hypothesized that early priming or sensitization of T cells may possibly begin during fetal life or the newborn period, and that the skewed Th_2 response to antigen stimulation may be exaggerated in individuals who have an atopic genotype (Rinas et al., 1993; Tang et al., 1994). Postnatally, a highly developed network of antigen presenting cells (dendritic cells) has been identified in the airway epithelium (Holt et al., 1997). Following exposure to relevant environmental stimuli (e.g. inhaled allergens and viral infections) these cells are candidates to play an important role at vulnerable time points in influencing the Th_1 vs Th_2 selection process. Other environmental exposures (e.g. passive smoke, indoor and outdoor air pollutants, and dietary factors) may also play a role in determining whether the development of Th_2 responsiveness is transient or persistent.

The key for developing new strategies for the treatment and prevention of asthma may be based on a better understanding for the relative importance of fetal, maternal, and post-natal events which influence the developing lung and immune system. The results of future studies in this area may also lead to new insights regarding the increasing incidence of asthma in Western societies. At present, too little is known to speculate whether a better understanding about the origins of asthma in early childhood will also improve our understanding for the development of asthma in those who begin having symptoms when they are teenagers or adults.

Acknowledgment

We are very grateful to Connie S. Lilly for her assistance in preparing this manuscript.

REFERENCES

Bjorksten, B., Epidemiology of pollution-induced airway disease in Scandinavia and Eastern Europe, *Allergy*, 52 (38 Suppl), 23, discussion 35, 1997.

Bjorksten, F., Suoniemei, and Kishi, V., Neonatal birch-pollen contact and subsequent allergy to birch pollen, *Clin. Allergy,* 10, 585, 1980.

Bousquet, J., and Kjellman, N.M., Predictive value of tests in childhood allergy, *J. Allergy Clin. Immunol.*, 78, 1019, 1986.

Brown M.A., Holonen, M.J., and Martinez, F.D., Cutting the cord: is birth already too late for primary prevention of allergy? *Clin. Exper. Allergy,* 27, 4, 1997.

Buffum, W.R., and Settipane, G.A., Prognosis of asthma in childhood. *Amer. J. Dis. Child.*, 112, 214, 1966.

Burney, P.G.J., Epidemiologic Trends, pp. 35-47, In: *Asthma*, Barnes, P.J., Grunstein, M.M., Leff, A.R., Woolcock, A.J. (Eds.), Lippincott-

Raven, Philadelphia, N.Y., 1997.

Burrows, B., Martinez, F.D., Halonen, M., Barbee, R.A., and Cline, M.G., Association of asthma with serum IgE levels and skin test reactivity to allergens, *N. Engl. J. Med.*, 320, 271, 1989.

Carosso, A., Ruffino, C., and Bugiani, M., The effect of birth season on pollenosis. *Ann. Allergy*, 56, 300, 1986.

Chandra, R.K., Puri, S., Cheema, and P.S., Predictive value of cord blood IgE in the development of atopic disease and the role of breast-feeding in its prevention, *Clin. Allergy*, 15, 517, 1985.

Chang, M.Y., Hogan, A.D., Rakes, G.P., Ingram, J.M., Hoover, G.E., Platts-Mills, T.A.E., and Heymann, P.W., Risk factors for high levels of salivary cotinine in children exposed to tobacco smoke. *J. Allergy Clin. Immunol.,* 95, 261, 1995.

Chin, J., Magoffin, R.L., Shearer, L.A., Schieble, J.H., Lennette, E.H., Field evaluation of a respiratory syncytial virus vaccine and trivalent parainfluenza virus vaccine in a pediatric population, *Am. J. Epidemiol.*, 89, 449, 1969.

Cookson, W.O.C.M, Sharp, P.A., Faux, J.A., and Hopkin, J.M., Linkage between immunoglobulin E responses underlying asthma and rhinitis and chromosome 11q. *Lancet,* I, 1292, 1989.

Croner, S., Kjellman, N.I.M., Development of atopic disease in relation to family history and cord blood IgE levels. Eleven-year follow-up in 1654 children, *Pediatr. Allergy Immunol.*, 1, 14, 1990.

De Alarcon, A., Walsh E., LaRussa, J.B., Rakes, G.P., Gleich, G., Platts-Mills, T.A.E., Carper, H.T., and Heymann, P.W., Antibody to respiratory syncytial virus (RSV) and eosinophil responses in wheezing infants, *Allergy Clin. Immunol. Int.*, Suppl. 4, 23, 1997.

Duff, A.L., Gelber, L.E., Deuell, B.L., Wilson, B.B., Farris, A.H., Platts-Mills, T.A.E., and Heymann, P.W., Development of inhalant allergy in early childhood: association with asthma and atopic dermatitis (AD). *J. Allergy Clin. Immunol.*, 89, 315, 1992.

Duff, A.L., Pomeranz, E.S., Gelber, L.E., Price, G.W., Farris, H., Hayden, F.G., Platts-Mills, T.A.E., and Heymann, P.W., Risk factors for acute wheezing in infants and children: viruses, passive smoke, and IgE antibodies to inhalent allergens, *Pediatrics*, 92, 535, 1993.

Frick, O.L., Role of viral infections in asthma and allergy. *Clin. Rev. Allergy*, 1, 5, 1983.

Garofalo, R., Kimpkin, J.L, Welliver, R.C., and Ogra, P.L., Eosinophil degranulation in the respiratory tract during naturally acquired respiratory syncytial virus infection. *J. Pediatr.*, 120, 28, 1992.

Gergen, P.J., Mullally, D.I., and Evans, R., National survey of prevalency of asthma among children in the United States, 1976 to 1980, *Pediatrics*, 81, 1, 1988.

Gergen, P.J., and Weiss, K.B., Changing patterns of asthma hospitalization among children, *J.A.M.A.*, 264, 1688, 1990.

Glezen, W.P., and Denny, F.W., Epidemiology of acute lower respiratory disease in children, *N., Engl. J. Med.*, 288, 498, 1973.

Guillet, G., and Guillet, M.H., Natural history of sensitizations in atopic dermatitis, *Arch. Dermatol.*, 128, 187, 1992.

Halonen, M., Stern, D., Taussig, L.M., Wright, A., Ray, C.G, Martinez, F.D., The predictive relationship between serum IgE levels at birth and subsequent incidences of lower respiratory illnesses and eczema in infants, *Am. Rev. Respir, Dis.*, 146, 866, 1992.

Heymann, P.W., Rakes, G.P., Hogan, A.D., Ingram, J.M, Hoover, G.E., and Platts-Mills, T.A.E., Assessment of eosinophils and IgE antibody in wheezing infants and children, *Int. Arch. Allergy Immunol.*, 107, 380, 1995.

Heymann, P.W., Zambrano, J.C., and Rakes G.P., Virus-induced wheezing in children: respiratory syncytial virus (RSV) and rhinovirus, *Immun. Clin. North Am.*, 18, 35, 1998.

Holt, P.G., Sly, P.D., and Björkstén, B., Atopic versus infectious diseases in childhood: a question of balance? *Pediatr. Allergy Immunol.*, 8, 53, 1997.

Ingram, J.M., Rakes, G.P., Hoover, G.E., Platts-Mills, T.A.E., and Heymann, P.W., Eosinophil cationic protein in serum and nasal washes from wheezing infants and children, *J. Pediatr.*, 127, 558, 1995.

Kemp, A.S., Relationship between the time of birth and the development of immediate hypersensitivity to grain pollen antigens. *Med. J. Aust.*, 1, 263, 1979.

Kim, H.W., Canchola, J.G, Brandt, C.D., Pyles, G., Chanock, R.M., Jensen, K., and Parrott, R.H., Respiratory syncytial virus disease in infants despite prior administration of antigenic inactivated vaccine, *Am. J. Epidemiol.*, 89, 422, 1969.

Koller, D.Y., Wojnarowski, C., Herkner, K.R., Weinländer, G., Raderer, M., Eichler I., and Frischer, T., High levels of eosinophil cationic protein in wheezing infants predict the development of asthma. *J. Allergy Clin. Immunol.* 99, 752, 1997.

Langeland, T., A clinical and immunological study of allergy to hen's egg white. *Clin. Allergy*, 13, 371, 1983.

La Russa, J.B., Rakes, G.P., Murphy, D.D., Platts-Mills, T.A.E., and Heymann, P.W., Assessments of eosinophil cationic protein (ECP) in nasal washes from wheezing infants, *J. Allergy Clin. Immunol.*, 97, 241, 1996.

LaVia, W.V., Mark, M.I., and Stuntman, H.R., Respiratory syncytial virus puzzle: clinical features, pathophysiology, treatment, and prevention, *J. Pediatr.*, 121, 503, 1992.

Lin, H., Mosmann, T.R., Guilbert, L., Tuntipopipat, S., and Wegmann, T.G., Synthesis of T helper 2-type cytokines at the maternal-fetal interface. *J. Immunol.* 151, 4562, 1993.

Marsh, D.G., Neeley, J.D., Breazeale, D.R., Ghosh, B., Freidhoff, L.R., Ehrlich-Kautzky, E., Schorr, C., Krishnaswamy, G., and Beaty, T.H., Linkage analysis of IL-4 and other chromosome 5q31.1 markers & total serum immunoglobulin E concentrations, *Science*, 264, 1152, 1994.

Martinez, F.D., Morgan, W.J., Wright A.L., Holber, C.J., Taussig L.M., and Group Health Medical Associates, Diminished lung function as a predisposing factor for wheezing respiratory illness in infants, *N. Engl. J. Med.,* 319, 1112, 1988.

Martinez, F.D., Stern, D.A., Wright, A.L., Holberg, C.J., Taussig, L.M., and Halonen, M., Association of interleukin-2 and interferon-gamma production by blood mononuclear cells in infancy with parental allergy skin tests and with subsequent development of atopy, *J. Allergy Clin. Immunol.*, 96, 652, 1995(a).

Martinez, F.D., Wright, A.L., Jansig, L.M., Holberg, C.J., Halonen, M., and Morgan, W.J., Asthma and wheezing in the first six years of life, *N. Engl. J. Med.*, 332, 133, 1995(b).

Miller, D.L, Hirvonen, T., and Gitlin, D., Synthesis of IgE by the human conceptus, *J. Allergy Clin. Immunol.*, 52, 182, 1973.

Morgan, W.J., and Martinez, F.D., Risk factors of developing wheezing and asthma in childhood, *Pediatr. Clin. North Am.*, 39, 1185, 1992.

Murray, A.B., and Morrison, B.J., The effect of cigarette smoke from the mother on bronchial responsiveness and severity of symptoms in children with asthma. *J. Allergy Clin, Immunol.*, 77, 575, 1986.

O'Keefe, E.B., Hochreuteuer, H.A., Levine, A.D., and Buckley, R.H., IgE synthesis by newborn B cells stimulated with recombinant IL-4 with and without hydrocortisone. *J. Allergy Clin. Immunol.*, 87, 209, 1991.

Openshaw, P.J., Clarke, S.L., and Record, F.M., Pulmonary eosinophilic response to respiratory syncytial virus infections in mice sensitized to the major surface glycoprotein G, *Int. Immunol.*, 4, 493, 1992.

Ownby, D.R., and McCullough, J., Passive exposure to cigarette smoke does not increase allergic sensitization in children. *J. Allergy Clin. Immunol.,* 82, 634, 1988.

Piccinni, M.P., Mecacci, F., Sampognaro, S., Manetti, R., Parronchi, P., Maggi, E., and Romagnani, S., Aeroallergen sensitization can occur during fetal life, *Int. Arch. Allergy Immunol.,* 102, 301, 1993.

Price, G.W., Hogan, A.D., Farris, A.H., Burks, A.W., Platts-Mills, T.A.E.,and Heymann, P.W., Sensitization (IgE antibody) to food allergens in wheezing infants and children, *J. Allergy Clin. Immunol.*, 96, 266, 1995.

Rinas, U., Horneff, G., Wahn, U., Interferon-γ production by cord-blood mononuclear cells is reduced in newborns with a family history of atopic disease and is independent from cord blood IgE-levels. *Pediatr.*

Allergy Immunol.,4, 60, 1993.

Rosenwasser, L.J., and Borish, L., Genetics of atopy and asthma: the rationale behind promoter-based candidate gene studies (IL-4 and IL-10), *Amer. J. Resp. Crit. Care Med.,* 156, S152, 1997.

Rowntree, S., Cogswell, J.J., Platts-Mills, T.A.E., and Mitchell, E.B., Development of IgE and IgG antibodies to food and inhalant allergens in children at risk of allergic disease, *Arch. Dis. Child.*, 75, 633, 1985.

Ruiz, R.G.G., Richards, D., Kemeny, D.M., and Price, J.F., Neonatal IgE: a poor screen for atopic disease, *Clin. Exp. Allergy*, 21, 467, 1991.

Shirakawa, T., Airong, L., Dubowitz, M., Dekker, J.W., Shaw, A.E., Faux, J.A., Ra, C., Cookson, W.O.C.M., and Hopkin, J.M., Association between atopy and variants of the beta subunit of the high-affinity immunoglobulin E receptor, *Nat. Genet.*, 7, 125, 1994.

Smith, J.M., and Springett, V.H., Atopic disease and month of birth. *Clin. Allergy,* 9, 153, 1979.

Sporik, R., Holgate, S.T., and Cogswell, J.J., Natural history of asthma in childhood - a birth cohort study, *Arch. Dis. Child.*, 66, 1050, 1991.

Sporik, R., Holgate, S.T., Platts-Mills, T.A.E., Cogswell, J.J., Exposure to house dust mite allergen *(Der p I)* and the development of asthma in childhood, *N. Engl. J. Med.,* 323, 502, 1990.

Srikiatkhachorn, A., and Braciale, T.J., Virus-specific memory and effector T lymphocytes exhibit different cytokine responses to antigens during experimental murine respiratory syncytial virus infection, *J. Virol.*, 71, 678, 1997.

Szépfalusi, Z.S., Nentwich, I., Gerstmayr, M., Jost, E., Todoran, L., Gratzl, R., Herkner, K., and Urbanek, R., Prenatal allergen contact with milk proteins, *Clin. Exper. Allergy*, 27, 28, 1997.

Tager, I.B., Hanrahan, J.P., Tostesen, T.D., Castile, R.G., Brown, R.W., Weiss, S.T., and Speizer, F.E., Lung function, pre-and post-natal smoke exposure, and wheezing in the first year of life. *Am. Rev. Respir. Dis.*, 147, 811, 1993.

Tang, M.L.K., Kemp, A.S., Thornburn, J., and Hill, D.J., Reduced interferon-γ secretion in neonates and subsequent atopy. *Lancet*, 344, 983, 1994.

Van Asperen, P.P., Kemp, A.S., and Mellis, C.M., A prospective study of the clinical manifestations of atopic disease in infancy, *Acta Pediatri. Scand.*, 73, 80, 1984.

Van der Ent, C.K., Smits, E.J.G.M., Schellekens, A.P.M., and Waelkens, J.J.J., Is IgE in umbilical cord blood really predictive? *Acta Pediatr. Scand.*, 79, 960, 1990.

Von Mutius, E., Martinez, F.D., Fritzsch, C., Nicolai, T., Roell, G., and Thiemann, H.H., Prevalence of asthma and atopy in two areas of West and East Germany, *Am. J. Respir, Crit. Care Med.*,149, 358, 1994.

Warner, J.A., Jones, A.C., Miles, E.A., Colwell, B.M., and Warner, J.O., Prenatal origins of asthma and allergy, *Ciba Foundation Symposium*, 206, 220, 1997.

Weil, G.J., Hussian, R., Kumaraswani, V., Tripathy, S.P., Phillips, K.S., and Ottesen, E.A., Prenatal allergic sensitization of helminth antigens in offspring of parasite-infected mothers. *J. Clin. Invest.*, 71, 1124, 1983.

Weitzman, M., Gortmaker, S.L., Sobol, A.M., and Perris, J.M., Recent trends in the prevalence and severity of childhood asthma. *JAMA*, 268, 2673, 1992.

Welliver, R.C., Wong, D.T., Sun, M, Middleton, E., Vaughan, R.S., and Ogra, P.L., The development of respiratory syncytial virus-specific IgE and the release of histamine in nasopharyngeal secretions after infection, *N. Engl. J. Med.*, 305, 841, 1981.

Welliver, R.C., Sun, M., Hildreth, S.W., Arumugham, R., and Ogra, P.L., Respiratory syncytial virus-specific antibody responses in immunoglobulin A and E isotypes to the F and G proteins and to intact virus after natural infection. *J. Clin. Microbiol.,* 27, 295, 1989.

Welliver, R.C., and Duffy, L., The relationship of RSV-specific immunoglobulin E antibody responses in infancy, recurrent wheezing, and pulmonary function at age 7-8 years, *Pediatr. Pulmonol.*, 15, 19, 1993.

Wilson, C.B., Westall, J., Johnston, L., Lewis, D.B., Dower, S.K., and Alpert, A.R., Decreased production of interferon-gamma by human neonatal cells. Intrinsic and regulatory deficiencies. *J. Clin. Invest.,* 77, 860, 1986.

Young, S., LeSouëf, P.N., Geelhoed, G.C., Stick, S.M., Chir, B., Turner, K.J., and Landau, L.I., The influence of a family history of asthma and parental smoking on airway responsiveness in infancy. *N. Engl. J. Med.*, 324, 1168, 1991.

Zeiger, R.S., and Heller, S., Development of nasal basophilic cells and nasal eosinophils from age 4 months through 4 years in children of atopic parents, *J. Allergy Clin. Immunol.,* 91, 723, 1993.

Zeiger, R.S., Heller, S., Mellon, M.H., Forsythe, A.B., O'Connor, R.D., Hamburger, R.N., and Schatz, M., Effect of combined maternal and infant food-allergen avoidance on development of atopy in early infancy: a randomized study. *J. Allergy Clin. Immunol.,* 84, 72, 1989.

Chapter 4

LOSS OF AIRWAY M_2 MUSCARINIC RECEPTOR FUNCTION FOLLOWING INFLAMMATION AND VIRAL INFECTION

David B. Jacoby, M.D. and Allison D. Fryer, Ph.D.

I. INTRODUCTION

Asthma affects over 10 million people in the United States (Schoenborn & Marano, 1989). The natural history of asthma includes periods of quiescence interrupted by exacerbations. These asthma attacks lead to hospitalization at some point for as many as 20% of patients with asthma.

Both inhalation of allergens and viral airway infections cause asthma attacks. In both cases, increased vagally-mediated reflex bronchoconstriction may be important. Our studies suggest that loss of function of an inhibitory receptor, the M_2 muscarinic receptor, on the vagus nerves may account for this increase in vagally-mediated bronchoconstriction.

II. INHIBITORY MUSCARINIC RECEPTORS ON PULMONARY PARASYMPATHETIC NERVES

The vagus nerves provide the dominant autonomic control of the lungs. In the airways, release of acetylcholine from postganglionic parasympathetic fibers stimulates smooth muscle contraction by binding to M_3 receptors on the smooth muscle. However, at the same time, released acetylcholine feeds back onto inhibitory M_2 receptors on the nerve endings themselves. Binding of acetylcholine to these inhibitory autoreceptors provides a negative feedback, decreasing further release of acetylcholine (Figure 1).

The importance of these inhibitory neuronal M_2 receptors becomes apparent when they are blocked using M_2 selective antagonists such as gallamine. By blocking feedback inhibition of acetylcholine release, gallamine can increase by as much as tenfold the bronchoconstrictor response to electrical stimulation of the vagus nerve. Conversely, stimulating the M_2 receptor with the muscarinic agonist pilocarpine can inhibit the bronchoconstrictor response to vagal stimulation by as much as 85%. These neuronal M_2 receptors are present in guinea pigs (Fryer & Maclagan, 1984), dogs (Ito & Yoshitomi, 1988), cats (Blaber *et al.*, 1985; Killingsworth *et al.*, 1992), and rats (Aas & Maclagan, 1990), and are also present in man (Minette & Barnes, 1988).

Inhibitory neuronal M_2 receptors are dysfunctional in guinea pigs after viral infections (Fryer & Jacoby, 1991), inhalation of allergens (Fryer & Jacoby, 1992; Fryer & Wills-Karp, 1991), and exposure to ozone (Schultheis *et al.*, 1994). The function of these receptors is also impaired

1-56670-373-5/99/$0.00+$.50

in some (Ayala & Ahmed, 1989; Minette *et al*, 1989), but not all (Okayama *et al.,* 1994), patients with asthma. All of these conditions are associated with increased parasympathetic drive to the airway smooth muscle (Boushey & Holtzman, 1985; Buckner *et al.,* 1985; McCaig, 1987). Loss of M_2 receptor function increases release of acetylcholine from the parasympathetic nerves, potentiating vagally-mediated bronchoconstriction, and thus may contribute to the airway hyperresponsiveness.

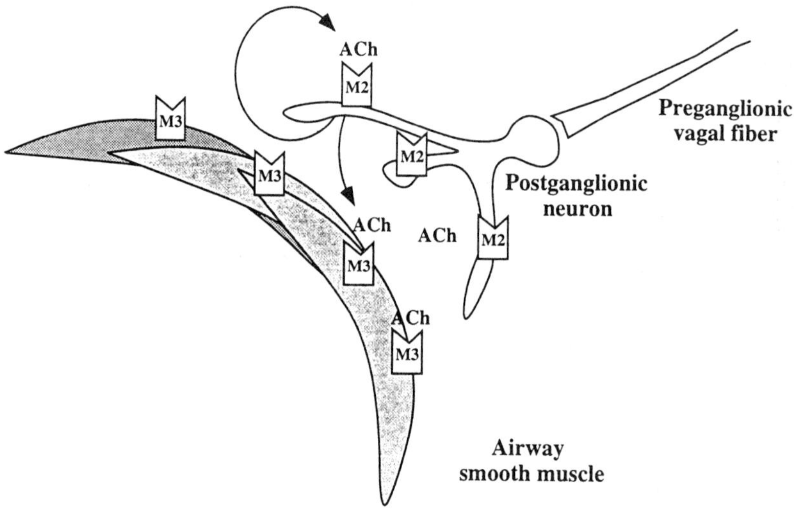

Figure 1. Acetylcholine is released from post-ganglionic vagal neurons in the airways. Binding of acetylcholine to M_3 muscarinic receptors on the airway smooth muscle causes broncho-constriction. At the same time, acetylcholine feeds back onto M_2 receptors on the nerve endings, turning off the further release of acetylcholine from the nerve.

The basic mechanisms for loss of M_2 receptor function have been studied in recent years, but are still incompletely understood. Both leukocyte-mediated and leukocyte-independent mechanisms are likely to participate. We have shown that enzymatic deglycosylation of M_2 receptors by parainfluenza virus neuraminidase decreases agonist affinity for the receptor by ten-fold (Fryer *et al.,* 1990). We have also shown that some inflammatory mediators, specifically eosinophil major basic protein and eosinophil peroxidase, are allosteric antagonists at the M_2 receptor (Jacoby *et al.,* 1993). These may also contribute to M_2 receptor dysfunction by acting as endogenous antagonists. Finally, the expression of the gene encoding the M_2 receptor may be decreased in disease (Jacoby *et al.,* 1998).

III. IMPORTANCE OF THE VAGUS IN ASTHMA

The predominant neural control of the airways is provided by the parasympathetic nervous system. While some studies have concluded that the cholinergic nerves are not important in asthma (Cockcroft *et al.*, 1978; Cockcroft *et al.*, 1977; Fish *et al.*, 1977; Rosenthal *et al.*, 1977; Ruffin *et al.*, 1978; Schiller & Lowell, 1947), these studies are flawed by the use of single, low doses of inhaled anticholinergic drugs without demonstrating adequate vagal nerve blockade. Anticholinergic agents appear to be poorly absorbed when given by inhalation. Thus, anticholinergic agents are effective bronchodilators when given intravenously, but are less effective in studies where they are given by inhalation (Holtzman *et al.*, 1983; Sheppard *et al.*, 1983). Furthermore, currently available anticholinergics are non-selective. Thus atropine and ipratropium may block M_2 receptors and increase acetylcholine release, overcoming any postjunctional blockade of the M_3 receptors on airway smooth muscles (Fryer & Maclagan, 1987). Despite these problems, high-dose atropine (Eiser & Guz, 1982; Itkin & Anand, 1970; Yu *et al.*, 1972) and ipratropium (Kresten, 1974; Orehek *et al.*, 1975; Schultze-Weringhaus, 1981) are effective bronchodilators.

Viral infection is also clearly associated with hyperresponsiveness that is vagally mediated (see below). Thus, despite some data suggesting that anticholinergics are not as effective as β-agonists as bronchodilators, there is solid evidence that the cholinergic nervous system does play an important role in many forms of asthma.

IV. INVOLVEMENT OF M_2 RECEPTOR DYSFUNCTION IN HYPERRESPONSIVENESS

We have demonstrated that the neuronal M_2 muscarinic receptors are dysfunctional in three different animal models of asthma and hyperresponsiveness, viral infection (Fryer & Jacoby, 1991), antigen challenge (Fryer & Jacoby, 1992) and exposure to ozone (Schultheis *et al.*, 1994). Loss of receptor function is a mechanism for airway hyperresponsiveness, as prevention of neuronal M_2 receptor dysfunction or restoration M_2 receptor function prevents or reverses hyperresponsiveness to vagal stimulation and to histamine (Evans *et al.*, 1997; Fryer & Jacoby, 1992) (Figure 2).

V. EFFECTS OF VIRAL INFECTIONS ON THE AIRWAYS

The strong epidemiological association of viral respiratory infections with asthma attacks has led to studies investigating the possible mechanisms of these effects. It is clear that there are changes in the neural control of the airways as shown by human studies of naturally-occurring infections (Aquilina *et al.*, 1980; Empey *et al.*, 1976). The mechanisms of neuronal hyperresponsiveness following virus, elucidated in animal studies, include increased response to tachykinins because of loss of neutral endopeptidase activity (Dusser *et al.*, 1989; Jacoby *et al.*, 1988) and increased vagally-mediated bronchoconstriction due to loss of inhibitory M_2 muscarinic receptors (Fryer & Jacoby, 1991) (see below). Viral infections may also decrease the function of inhibitory β-adrenoceptors on leukocytes (Buckner *et al.*, 1981; Busse, 1977; Busse *et al.*, 1980; Busse *et al.*, 1979), activate leukocytes (Busse *et al.*, 1991), augment epithelial cell cytokine production (Choi & Jacoby, 1992; Einarsson *et al.*, 1996; Noah & Becker, 1993; Subauste *et al.*, 1995), damage epithelial cells (Jacoby *et al.*, 1988; Walsh *et al.*, 1960), and enhance the late-phase asthmatic response to inhaled allergens (Lemanske *et al.*, 1989). Furthermore, in allergic animals, virus-stimulated CD8+ T-lymphocytes, which normally produce interferon-γ, can produce interleukin-5, an eosinophil chemotaxin and activator (Coyle *et al.*, 1995). Thus there is potential for viral infections to interact with antigens to worsen asthma.

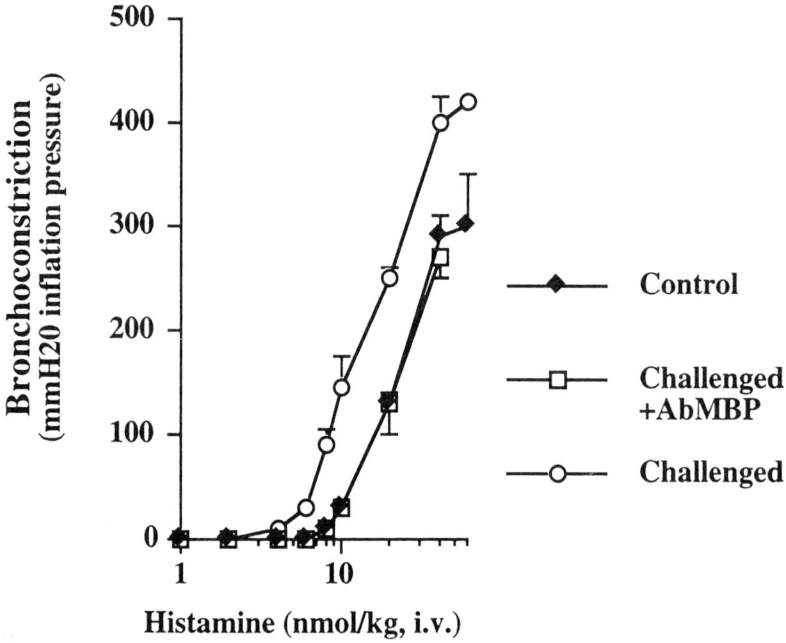

Figure 2. Inhalation of allergen by sensitized guinea pigs causes M_2 receptor dysfunction and hyperresponsiveness to histamine, an agonist that is known to have a vagally-mediated reflex component. When M_2 receptor dysfunction is prevented by pretreating animals with an antibody to eosinophil major basic protein, histamine hyperresponsiveness is also prevented (Evans et al., 1997).

VI. VIRAL INFECTIONS IN ASTHMA

The role of viral infections in causing asthma attacks has been a topic of considerable interest in recent years. Viral respiratory infections frequently exacerbate asthma, both in children (Frick *et al.,* 1979; Henderson *et al.,* 1979; Welliver, 1983) and in adults (Little *et al.,* 1978; Welliver, 1983). A recent study by Johnston and colleagues found evidence of viral infections in 80% of exacerbations of asthma in a group of children followed for two years (Johnson *et al.,* 1995) (Figure 3). The finding of temporary airway hyperresponsiveness following acute viral infections in previously normal subjects (Aquilina *et al.,* 1980; Empey *et al.,* 1976; Little *et al.,* 1978) suggests that viral infection may be important in the pathogenesis of asthma and, particularly, in asthma attacks.

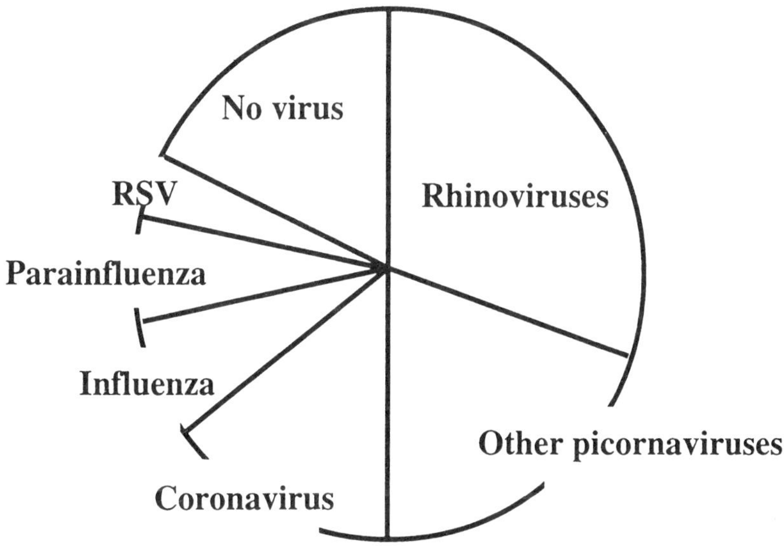

Figure 3. Viral infections were detected in 80% of asthma attacks in a group of children ages 9-11 (Johnston *et al.*, 1995).

A. EFFECTS OF ACUTE VIRAL INFECTION ON M_2 RECEPTOR FUNCTION *IN VIVO*

Our initial studies in this area (Fryer & Jacoby, 1991) were done in guinea pigs that had been infected with parainfluenza type 3. Four days after infection, guinea pigs were anaesthetized, tracheostomized, and mechanically ventilated. The vagus nerves were cut and the distal portions were stimulated electrically. Bronchoconstriction was measured as an increase in pulmonary inflation pressure given constant volume and flow.

In normal guinea pigs vagal stimulation caused a reproducible bronchoconstriction that was potentiated by the M_2 antagonist gallamine and inhibited by the muscarinic agonist pilocarpine. In contrast, in virus-infected animals vagally-induced bronchoconstriction was increased, and pilocarpine did not inhibit vagally-mediated bronchoconstriction. Furthermore, the potentiation caused by gallamine was markedly less in virus-infected animals, even after adjusting the stimulating voltage so that the baseline response to stimulation was equivalent in the virus-infected and control animals. Thus, viral infection caused a marked loss of function of inhibitory neural M_2 muscarinic receptors.

B. ROLE OF VIRAL NEURAMINIDASE

The M_2 receptor is N-glycosylated at three asparagine residues (Asn^2, Asn^3, and Asn^6) in the extracellular portion of the receptor. Carbohydrates make up 26.5% of the molecular weight of the receptor (Peterson *et al.*, 1986). Gies and Landry (Gies & Landry, 1988) showed that treatment of M_2 receptors with enzyme neuraminidase, which cleaves the sialic acid residues from the receptor, decreases the affinity of muscarinic agonists for the M_2 receptor but not for other muscarinic receptor subtypes.

Neuraminidase is contained in the coat of both influenza and parainfluenza viruses (Scheid *et al.*, 1972) and is expressed in large quantities in infected tissues (Boulan & Pendergast, 1980). To determine whether parainfluenza virus could directly affect muscarinic receptor function, we performed ligand-receptor binding studies M_2 receptors incubated *in vitro* with parainfluenza virus or with neuraminidase. These treatments did not affect total receptor number. However, agonist affinity for the M_2 receptor was decreased by an order of magnitude. The addition of a neuraminidase inhibitor prevented virus-induced loss of agonist affinity.

C. THE ROLE OF INFLAMMATION IN VIRUS-INDUCED LOSS OF M_2 RECEPTOR FUNCTION

Because we have shown that the inflammatory responses to antigen (Elbon *et al.*, 1995) and to ozone (Gambone *et al.*, 1994) are the cause of the loss of M_2 receptor function, we studied the role of inflammation in virus-infected animals. We pretreated guinea pigs with cyclophosphamide (30 mg/kg) daily for 7 days, after which they were infected with parainfluenza virus. Four days later their response to vagal stimulation and to pilocarpine and gallamine were studied. Of ten guinea pigs studied, cyclophosphamide protected M_2 receptor function in four animals despite viral infection. The remaining six lost M_2 receptor function despite being depleted of leukocytes as determined by both peripheral blood and lung lavage cell counts.

Because of previous reports suggesting that virus may replicate to higher levels in the lungs of leukocyte-depleted animals (Hurd & Heath, 1975;

Singer *et al.*, 1972), we determined the viral content of the lungs in this experiment. Guinea pigs that lost M_2 receptor function despite leukocyte depletion had significantly and substantially greater lung viral content than those that retained normal M_2 receptor function. To examine further the relationship between the severity of viral infection and the effect on M_2 receptor function (in the absence of inflammation), we compared the viral content of the lungs with the response to vagal stimulation observed after the maximal dose of pilocarpine (100 μg/kg, i.v.). A highly significant correlation of viral content to M_2 receptor dysfunction was present.

Thus viral infection causes loss of M_2 receptor function by causing inflammation, as seen in animals with mild infections (low viral titers). There is also a leukocyte-independent effect that is seen in animals with severe viral infections (high viral titers). The mechanism of this effect may be cleavage of sialic acid residues from M_2 receptors by viral neuraminidase (as we described with ligand binding studies above), although this has not been directly demonstrated *in vivo*.

D. INFLAMMATORY RESPONSES TO VIRAL INFECTIONS

Inflammation, while not the only mechanism, does participate in virus-induced M_2 receptor dysfunction. Airway viral infections have been traditionally thought to cause a neutrophil influx into the airways followed by a mononuclear cell influx (Walsh *et al.*, 1960), and to elicit a predominantly CD8+ T-lymphocyte response (Ada *et al.*, 1981) yielding interferon-γ as an important anti-viral cytokine (Samuel, 1991). It has become apparent, however, that viral infections are capable of producing a varied inflammatory response, including airway eosinophilia, CD4+ T-lymphocyte responses, and production of interleukin-5 by CD8+ T-lymphocytes.

Eosinophils and their proteins have been recovered from both humans and mice with respiratory syncytial virus infections (Garofalo *et al.*, 1992; Openshaw *et al.*, 1992; Sigurs *et al.*, 1994). In allergic humans, experimental rhinovirus infections increase the recruitment of eosinophils to the airways after antigen challenge (Calhoun *et al.*, 1994). In view of the association of eosinophils with asthma and asthma exacerbations (Frigas *et al.*, 1981), this eosinophil influx may be an important mechanism of virus-induced asthma. Furthermore, given the effects of eosinophil proteins (particularly major basic protein) as M_2 receptor antagonists, virus-induced eosinophil influx may contribute to loss of M_2 receptor functions. The recent demonstration that an anti-major basic protein antibody decreases antigen-induced hyperresponsiveness *in vivo* (Lefort *et al.*, 1996) provides another tool to explore the role of major basic protein in M_2 receptor dysfunction.

The mechanisms of virus-induced eosinophilia of the airways are not known. We (Choi & Jacoby, 1992; Subauste *et al.*, 1995), and others

(Noah & Becker, 1993), have shown that airway epithelium releases interleukin-8 in response to viral infections. Interleukin-8 is chemotactic for eosinophils, particularly after they are primed by interleukin-5 (Douglass *et al.*, 1994; Schweizer *et al.*, 1994). Interleukin-5, which both attracts and activates eosinophils, is a product of Th2 lymphocytes and of eosinophils themselves. However, it has recently been shown that while under normal circumstances CD8+ T-lymphocytes produce interferon-γ in response to viral infections, in an allergic milieu these CD8+ cells respond to viral infections by producing interleukin-5 (Baumgarth *et al.*, 1994; Coyle *et al.*, 1995). This, in addition to the realization of the importance of the virus-induced CD4+ T-lymphocyte response (Scherle *et al.*, 1992), will likely contribute to our understanding of virus-induced airway eosinophilia and asthma attacks. Furthermore, infection with a variety of viruses, including parainfluenza, increases endothelial expression of VCAM-1 (Christensen *et al.*, 1995; Smolinski *et al.*, 1995; Viac *et al.*, 1995), the endothelial receptor for VLA-4, which is an important adhesion molecule for both eosinophil and lymphocyte migration (Bochner *et al.*, 1991; Schleimer *et al.*, 1992).

Viral infections also promote and potentiate neurogenic inflammation of the airways (Piedimonte *et al.*, 1989) by decreasing neutral endopeptidase, an enzyme that degrades tachykinins. In view of the recently recognized importance of tachykinins in the inflammatory response to inhaled antigen (Bertrand *et al.*, 1993), potentiation of neurogenic inflammation may also contribute to heightened allergen responses in virus-infected airways. We have shown that neurokinin receptor antagonists block antigen-induced loss of M_2 receptor function.

VII. SYSTEMS FOR THE STUDY OF THE REGULATION OF M_2 RECEPTOR GENE EXPRESSION

Study of the regulation of the expression of the M_2 receptor has been stymied by the fact that the promoter for this gene has not been cloned. The genes for all five muscarinic receptors contain splice sites in the 5' untranslated region just upstream from the open reading frame. Study of the control of the rat M_4 receptor gene has recently been begun using a 25000 base upstream segment (Wood *et al.*, 1995). No such studies have been done with a mammalian M_2 receptor, although the 5'-flanking region for the chicken M_2 receptor has recently been cloned (Rosoff *et al.*, 1996). Both the rat M_4 receptor and the chicken M_2 receptor genes have large introns just upstream from the open reading frame.

Several systems have been used to try to study M_2 receptor gene expression. Chick heart fibroblasts express muscarinic receptors, and their regulation and function has been studied (Hamilton *et al.*, 1995). An alternative system utilizes the HEL 299 human lung fibroblast cell line, which express muscarinic receptors of the M_2 subtype only. In these cells expression of the M_2 receptor gene is down-regulated after treatment of the cells with phorbol ester (Rousell *et al.*, 1995) or with transforming growth factor-β (Haddad *et al.*, 1996). This group has continued to study this system to try to define the signal transduction pathway of the regulation M_2 receptor expression in these cells. They have recently suggested that activation of the transcriptional regulatory protein NF-κB may be involved (Haddad *et al.*, 1996), and that glucocorticoids may prevent down-regulation of the expression of this receptor (Haddad *et al.*, 1995).

VIII. MUSCARINIC RECEPTORS ON CULTURED AIRWAY PARASYMPATHETIC NEURONS & THE EFFECTS OF VIRAL INFECTION

A. CHARACTERIZATION OF AIRWAY PARASYMPATHETIC NERVE CELLS IN CULTURE

To complement our *in vivo* functional studies and our *in vitro* receptor-ligand binding studies, we developed a method for growing airway parasympathetic neurons in culture (Fryer *et al.*, 1996). These neurons synthesize and release acetylcholine, and they express M_2 muscarinic receptors that limit the release of acetylcholine (Figure 4). The presence of other muscarinic receptor subtypes in these cells was excluded by both immunostaining and reverse transcription-polymerase chain reaction analysis.

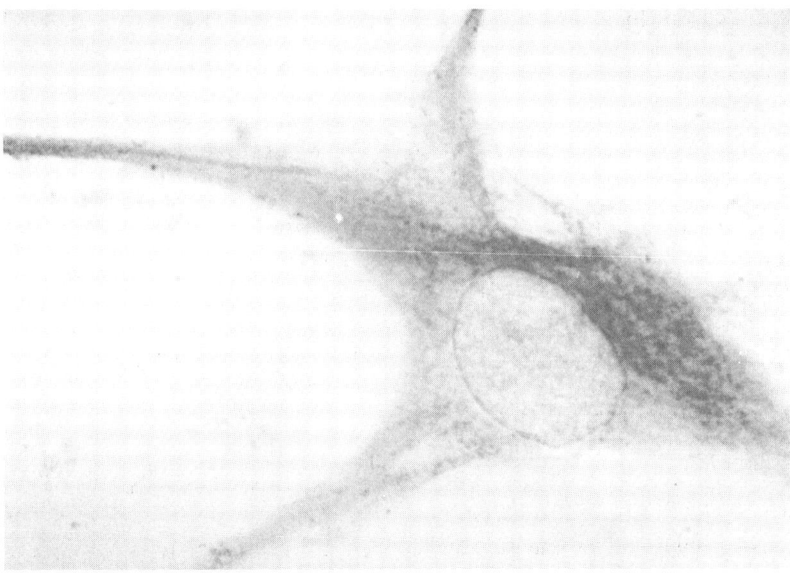

Figure 4. Airway parasympathetic neuron in culture stained with an antibody to the M_2 muscarinic receptor (Fryer *et al.*, 1996). Staining with antibodies to other muscarinic receptor subtypes was negative.

By blocking the M_2 receptors on these nerve cells, atropine potentiates the release of acetylcholine. Conversely, methacholine stimulates these receptors and decreases the release of acetylcholine.

B. VIRAL INFECTION OF CULTURED NERVE CELLS

We have also shown that these cells will support the growth of parainfluenza virus, allowing study of the direct interaction of viruses and neurons. We have measured acetylcholine release from virus infected nerve cell cultures. Viral infection markedly potentiated the release of acetylcholine in response to electrical stimulation. In virus-infected cells, the release of acetylcholine was neither potentiated by atropine nor suppressed by methacholine, demonstrating loss of M_2 receptor function. Similar results followed treatment of uninfected cells with interferon-γ, a cytokine produced by lymphocytes in response to viral infections.

In order to determine whether these effects of viral infections and interferon-γ were the result of a decrease in the expression of the gene encoding the M_2 receptor, we developed a competitive reverse transcription-polymerase chain reaction assay. Using this assay, we were able to demonstrate greater than an order of magnitude decrease in M_2 receptor mRNA in both virus-infected and interferon-γ-treated cells.

Thus both viral infection and interferon-γ substantially reduce nerve cell M_2 receptor mRNA levels. The association of this finding with clear evidence of loss of M_2 receptor function suggests that one possible explanation is loss of receptor synthesis because of decreased gene expression.

IX. ACUTE EFFECTS OF ANTIGEN CHALLENGE ON M_2 RECEPTOR FUNCTION

Dysfunction of the M_2 receptor also causes hyperresponsiveness to vagal stimulation in antigen-challenged guinea pigs (Fryer & Wills-Karp, 1991). This may be relevant to the study of virus-induced M_2 receptor dysfunction in that: (1) both stimuli produce an inflammatory response that includes the airway nerves; and (2) viral infections may interact with allergy, potentiating the allergic response to antigen and producing eosinophilia even in the absence of allergen.

A. THE ROLE OF THE EOSINOPHIL IN ANTIGEN-INDUCED M_2 RECEPTOR DYSFUNCTION

Because of the intense eosinophilic infiltrate in the lungs of antigen-challenged animals, we hypothesized that the eosinophil might be involved in the loss of M_2 receptor function (Fryer & Jacoby, 1992). Specifically, it has been observed that many M_2 receptor antagonists are positively charged, and it has been suggested that this may be a critical requirement in the binding of a ligand to the M_2 receptor (Hu *et al.*, 1992). Three of the granular proteins of the eosinophil, major basic protein, eosinophil cationic protein, and eosinophil peroxidase, are among the most strongly positively charged mammalian proteins. We postulated that these positively charged proteins might be functioning as endogenous antagonists at the M_2 receptor in conditions associated with eosinophilic inflammation.

We carried out three series of experiments to test this hypothesis. To block the eosinophil influx into the lungs, we pre-treated guinea pigs with monoclonal antibodies to either interleukin-5 or VLA-4. To attempt to reverse the effects of positively charged proteins on the M_2 receptor *in vivo*, we studied the effects of the poly-anionic substances heparin (a polysaccharide) and poly-l-glutamic acid (a polypeptide) on M_2 receptor function in control and antigen-challenged guinea pigs (Fryer & Jacoby, 1992). Both of these substances are known to neutralize positively charged eosinophil proteins *in vitro* (Barker *et al.*, 1989; Frigas *et al.*, 1980). We also directly measured the affinity of eosinophil proteins for M_2 receptors *in vitro* (Jacoby *et al.*, 1993).

Effects of blocking eosinophil influx: It has been observed that pretreatment with an antibody to interleukin-5 blocks the eosinophil influx into the airways of experimental animals after antigen challenge (Chand *et al.*, 1992; Gulbenkian *et al.*, 1992). We treated ovalbumin-sensitized guinea pigs with TRFK-5 (a monoclonal antibody to interleukin-5) before exposing them to aerosols of ovalbumin as described above. The antibody blocked the influx of eosinophils into the lungs as assessed both histologically and by whole lung lavage (Elbon *et al.*, 1995). Likewise, although antigen challenge caused loss of M_2 receptor function in untreated guinea pigs, in those guinea pigs that received the interleukin-5 antibody, normal responses to pilocarpine and gallamine were observed, demonstrating protection of M_2 receptor function.

Another molecule that is important in antigen-induced lung eosinophilia is VLA-4, an adhesion molecule on the eosinophil that binds to VCAM-1 on endothelial cells. We used HP1/2, a monoclonal antibody to VLA-4, to attempt to block the effects of antigen challenge on the M_2 receptor. Again, we were able to block the influx of eosinophils, and the loss of M_2 receptor function was avoided.

Because of reports that tachykinins may play an important role in airway inflammation after antigen inhalation (Bertrand *et al.*, 1993), we also studied the effects of neurokinin receptor antagonists. Pretreating sensitized guinea pigs with CP-96,345 (an NK-1 receptor antagonist) before antigen challenge prevented loss of M_2 receptor function. However, in this case the influx of eosinophils is not prevented. Our preliminary studies into the mechanisms by which NK-1 receptor antagonists prevent antigen-mediated loss of M_2 receptor function suggest that tachykinins may be involved in eosinophil activation.

The role of eosinophil proteins: Knowing that receptor glycosylation, particularly with the negatively charged sialic acid, was important to M_2 receptor function, and that many M_2 receptor antagonists are positively charged (Hu *et al.*, 1992), and that the eosinophil appears to be crucial to the effects of antigen on M_2 receptor function, we reasoned that one of the positively-charged proteins of the eosinophil might be acting as an endogenous M_2 receptor antagonist. We used both *in vivo* studies, in which we attempted to neutralized these positively-charged proteins, and receptor-ligand binding studies, to investigate this possibility.

Neutralizing positively charged eosinophil proteins: Our initial studies (Fryer & Jacoby, 1992) took advantage of the fact that heparin and poly-1-glutamic acid bind and neutralize positively charged eosinophil proteins. We determined the response to vagal stimulation, as above, in antigen-challenged guinea pigs. After a stable response to vagal stimulation was obtained, either heparin (2000 U/kg, i.v.) or poly-1-glutamic acid (15 mg/kg, i.v.) was given and the responses to vagal stimulation were measured every minute thereafter. The response to vagal stimulation was unchanged for 5 minutes, after which a progressive decrease in vagally-mediated bronchoconstriction was observed over 20 min, reaching a plateau at approximately 50% of the bronchoconstriction before administration of the heparin or poly-1-glutamic acid (Figure 5).

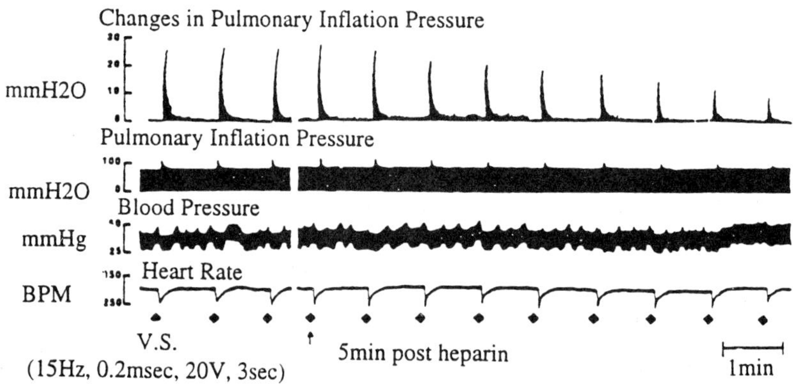

Figure 5. Treating an antigen-challenged guinea pig with heparin (2000 U/kg, i.v.) reverses M_2 receptor dysfunction acutely. Return of the negative feedback inhibition of acetylcholine release is seen as a decrease in the bronchoconstriction response to vagal stimulation (Fryer & Jacoby, 1992).

Responses to intravenous acetylcholine were unchanged by heparin or poly-1-glutamic acid, demonstrating that the M_3 receptors on the airway smooth muscle were not affected. Neither of the poly-anionic substances affected vagally-induced or intravenous acetylcholine-induced bronchonconstriction in control animals. After a new stable baseline was achieved, we tested the ability of pilocarpine to suppress, and of gallamine to potentiate, vagally-induced bronchoconstriction. Treatment with heparin or poly-1-glutamic acid, in addition to decreasing the bronchoconstrictor response to vagal stimulation, also restored the effects of pilocarpine and of gallamine on vagally-induce bronchoconstriction.

Thus the decrease in vagally-mediated bronchonstriction after heparin or poly-1-glutamate is accompanied by, and probably due to, a restoration of function of the M_2 receptor. The effect of these two substances, structurally very different but both poly-anionic and both capable of neutralizing eosinophil proteins *in vitro* (Gundel *et al.*, 1991), is consistent with the positively charged proteins of the eosinophil being responsible for the loss of M_2 receptor function.

We have recently demonstrated that pretreating guinea pigs with an antibody to eosinophil major basic protein before inhalation of allergen prevents loss of M_2 receptor function (Evans *et al.,* 1997). Thus vagally-mediated hyperresponsiveness after allergen challenge is mediated by major basic protein.

Radioligand binding studies of eosinophil proteins: Because negatively charged substances such as heparin and poly-1-glutamic acid restore M_2 receptor function, we examined whether endogenous positively charged proteins such as eosinophil major basic protein are antagonists for the M_2 receptor. We undertook a series of *in vitro* ligand binding studies using purified eosinophil major basic protein, eosinophil cationic protein, eosinophil peroxidase, and eosinophil-derived neurotoxin. Using membrane preparations of guinea pig heart (M_2 receptors) and guinea pig submandibular gland (M_3 receptors) to study the effects of these proteins on ligand binding, we found that major basic protein displaced ^{3}H-N-methylscopalamine from heart membranes (M_2), but not from gland membranes (M_3) (Jacoby *et al.,* 1993), with a dissociation constant of about 1.5 X 10^{-5}M (Figure 6). This is within the likely concentration range for this protein in the airways (Frigas *et al.*, 1981), and is therefore likely to be physiologically relevant.

Similar effects were demonstrated with eosinophil peroxidase, although this protein was less potent than major basic protein. Interestingly, eosinophil cationic protein did not share this property, demonstrating that structural determinants, as yet undefined, other than positive charge are required for a positively charged protein to be an M_2 antagonist. We have also recently shown similar effects of major basic proteins on human

M_2 muscarinic receptors using COS-7 cells stably transfected with each of the five human subtypes (Jacoby *et al.*, 1995).

The major basic protein is a selective endogenous antagonist at the M_2 receptor, a finding that may explain antigen-induced vagal hyperresponsiveness, and which may contribute to heightened vagally-mediated bronchoconstrictor responses after ozone exposure or viral infection.

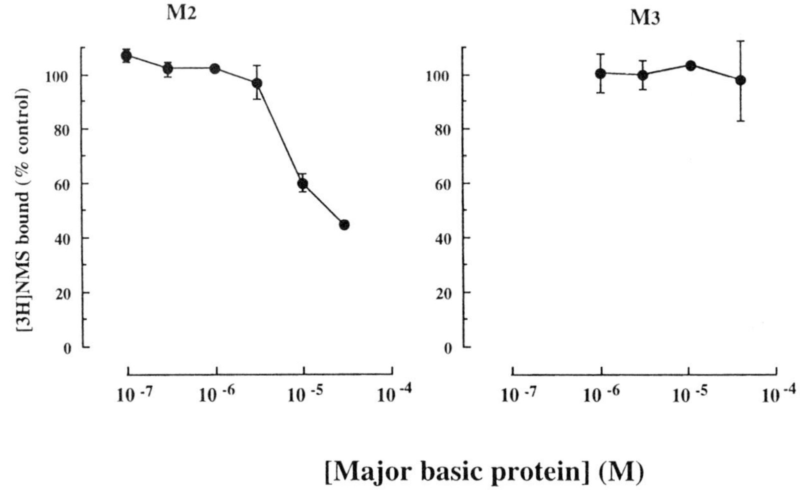

Figure 6. Human eosinophil major basic protein binds to M_2 receptors, displacing 3H-N-methylscopolamine. In contrast, there is no binding to M_3 receptors (Jacoby *et al.*, 1993).

B. EOSINOPHILIC INFLAMMATION OF AIRWAY NERVES

Further credence is lent to the hypothesis that the eosinophil is involved in loss of M_2 receptor function after antigen challenge by our recent finding that histological sections of the airways of antigen-challenged guinea pigs reveal eosinophils surrounding and infiltrating the nerves (Figure 7).

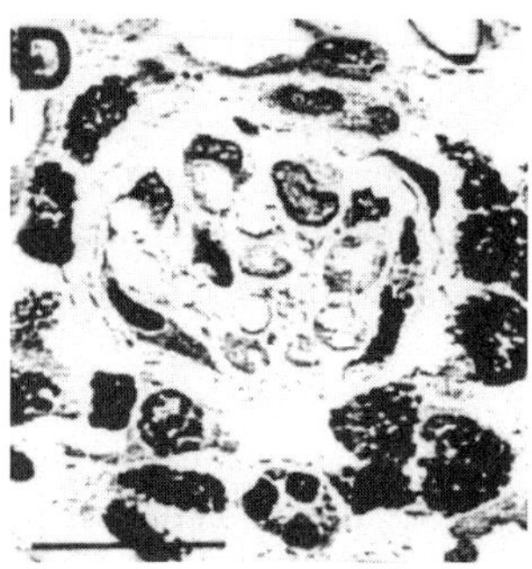

Figure 7. Airway nerve from antigen-challenged guinea pig, seen in cross section, surrounded by eosinophils (Costello *et al.*, 1997).

Using a computer image analysis system, we counted the number of eosinophils per airway nerve in antigen challenged and control animals in which the M_2 receptor function had been tested *in vivo* using pilocarpine. The number of eosinophils correlated with the M_2 receptor function expressed as the suppression of the response to bronchoconstriction after the maximal dose of pilocarpine (100 μg/kg). There was no correlation of M_2 receptor function with either the number of macrophages or the number of lymphocytes in the nerves. Furthermore, the eosinophils appeared to cluster around the nerves, being found in about twice the density around the nerves as they were around the adjacent blood vessels, and substantially higher than in other parts of the airway (Costello *et al.*, 1997).

We have also demonstrated that eosinophils are associated with nerves in the airways of people who have died with acute severe asthma (Figure 8).

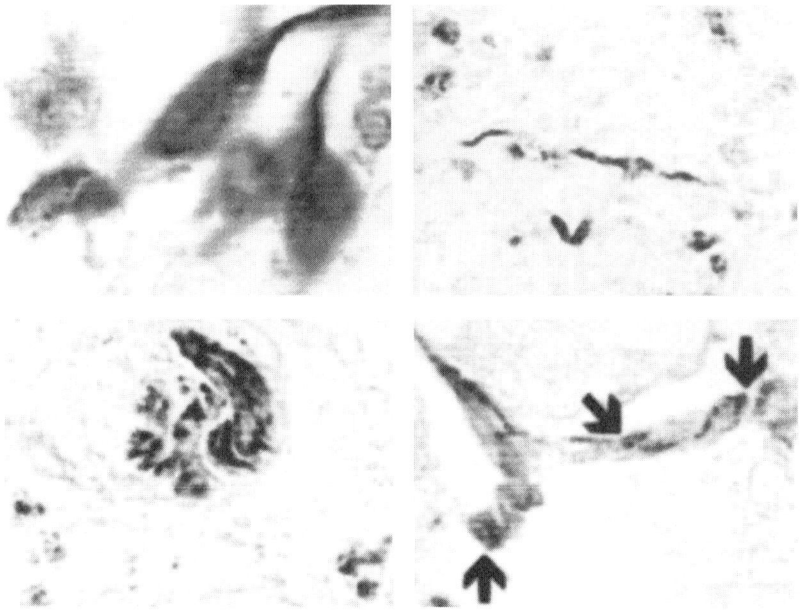

Figure 8. Hematoxylin and eosin staining of airway sections from patients who died of asthma. UPPER LEFT: Airway parasympathetic ganglion, showing two nerve cell bodies in close association with two eosinophils. LOWER LEFT: Cross section of nerve bundle showing an eosinophil within the bundle. UPPER RIGHT: Longitudinal section of nerve fiber within airway smooth muscle, showing an eosinophil adherent to nerve fiber. LOWER RIGHT: Section stained with antibody to human eosinophil major basic protein, with detection system showing major basic protein. Arrows show extracellular major basic protein deposited along airway nerve fiber (Costello *et al.*, 1997).

Thus eosinophils are recruited to the airway nerves in antigen-challenged guinea pigs and in humans with asthma. Release of eosinophil major basic protein blocks binding of acetylcholine to M_2 receptors on the nerves, eliminating the negative feedback control of acetylcholine release and causing vagally-medicated hyperresponsiveness.

X. CONCLUSION

Thus, loss of function of inhibitory M_2 muscarinic receptors on the airway parasympathetic nerves causes vagally-mediated bronchoconstriction and hyperresponsiveness after viral infections and allergen inhalation. Allergen-induced M_2 receptor dysfunction is absolutely dependent upon an influx of eosinophils into the airways. Activated eosinophils release major basic protein, which binds to M_2 receptors and prevents binding of acetylcholine. Thus the normal negative feedback control of acetylcholine release is lost, and acetylcholine release is increased.

In contrast, eosinophils are not involved in M_2 receptor dysfunction and hyperresponsiveness after viral infections. In this setting, both inflammatory cell products (possibly interferon-γ) and direct effects of viruses may be responsible. Deglycosylation of M_2 receptors by viral neuraminidase decreases acetylcholine affinity, again eliminating the negative feedback control of acetylcholine release. Furthermore, both parainfluenza virus and interferon-γ substantially decrease the expression of the gene encoding the M_2 receptor in airway parasympathetic nerves. Thus decreased synthesis of M_2 receptors may also be involved.

In atopic individuals, however, the inflammatory response to viral infections may be altered. In such individuals airway eosinophilia may occur. Furthermore, viral infections may potentiate the response to subsequent allergen inhalation. In such settings it is possible that the mechanisms of M_2 receptor dysfunction may more closely approximate those seen after allergen challenge.

Thus there are multiple mechanisms by which airway diseases can impair M_2 receptor function and cause vagally-mediated hyperresponsiveness. Eludication of these mechanisms will allow recognition of targets for specific therapies to prevent or reverse these causes of asthma attacks.

REFERENCES

Aas, P. & Maclagan, J., Evidence for prejunctional M_2 muscarinic receptors in pulmonary cholinergic nerves of the rat. *Br. J. Pharmacol.* 101, 73-76, 1990.

Ada, G.L., Leung, K.N. & Ertl, H., An analysis of effector T cell generation and function in mice exposed to influenza A or Sendai viruses. *Immunological Rev.* 58, 5-24, 1981.

Aquilina, A.T., Hall, W.J. Douglas, R.G. & Utell, M.J., Airway reactivity in subjects with viral upper respiratory tract infections: the effects of exercise and cold air. *Am. Rev. Respir. Dis.* 122, 3-10, 1980.

Ayala, L.E. & Ahmed, T., Is there loss of a protective muscarinic receptor in asthma? *Chest* 96, 1285-1291, 1989.

Barker, R.L., Loegering, D.A., Ten, R.M., Hamman, K.L., Pease, R.L. & Gleich, G.J., Eosinophil cationic protein cDNA: comparison with other toxic cationic proteins and ribonucleases. *J. Immunol.* 143, 952-955, 1989.

Baumgarth, N., Brown, L., Jackson D. & Kelso, A., Novel features of the respiratory tract T cell response to influenza virus infection: lung T cells increase expression of gamma interferon mRNA *in vivo* and maintain high levels of mRNA expression for interleukin-5 (IL-5) and IL-10. *J. Virol.* 68, 7575-81, 1994.

Bertrand, C., Geppetti, P., Baker, J., Yamawaki, I. & Nadel, J.A., Role of neurogenic inflammation in antigen-induced vascular extravasation in guinea pig trache. *J. Immunol.* 150, 1479-85, 1993.

Blaber, L.C., Fryer, A.D. & Maclagan, J., Neuronal muscarinic receptors attenuate vagally-induced contraction of feline bronchial smooth muscle. *Br. J. Pharmacol.* 86, 723-728, 1985.

Bochner, B.S., Lucinskas, F.W., Gimbrone, M.A., Newman, W., Sterbinsky, S.A., Derse-Anthony, C.P., Klunk, D. & Schleimer, R.P., Comparison of endothelial cell adhesion molecules for human basophils, eosinophils, and neutrophilis. *J. Exp. Med.* 173, 1553-1557, 1991.

Boulan, E.R. & Pendergast, M., Polarized distribution of viral envelope proteins in the plasma membrane of infected epithelial cells. *Cell* 20, 45-54, 1980.

Boushey, H. & Holtzman, M., Experimental airway inflammation and hyperreactivity; searching for cells and mediators. *Am. Rev. Respir. Dis.* 131, 312-313, 1985.

Buckner, C.K., Clayton, D.E., Ain-Shoka, A.A., Busse, W.W., Dick, E.C. & Shult, P., Parainfluenza 3 infection blocks the ability of a beta adrenergic receptor to inhibit antigen-induced contraction of guinea pig isolated airway smooth muscle. *J. Clin. Invest.* 67, 376-384, 1981.

Buckner, C.K., Songsirdej, V., Dick, E. C. & Busse, W.W., *In vivo* and *in vitro* studies of the use of the guinea pig as a model for virus-provoked airway hyperreactivity. *Am. Rev. Respir. Dis.* 132, 305-310, 1985.

Busse, W.W., Decreased granulocyte response to isoproterenol in asthma during upper respiratory infections. *Am. Rev. Respir. Dis.* 115, 783-791, 1977.

Busse, W.W., Anderson, C.L., Dick, E.C., & Warshauer, D.M., Reduced granulocyte response to isoproterenol, histamine, and prostaglandin E1 after *in vitro* incubation with rhinovirus 16. *Am. Rev. Respir. Dis.* 122, 641-646, 1980.

Busse, W.W., Cooper, W., Warshauer, D.M., Dick, E.C., Wallow, I.H.L. & Albrecht, R., Impairment of isoproterenol, H2 histamine, and prostaglandin E1 response of human granulocytes after incubation *in vitro* with live influenza vaccines. *Am. Rev. Respir. Dis.* 119, 561-569, 1979.

Busse, W.W., Vrtis, R.F., Steiner, R. & Dick, E.C., *In vitro* incubation with influenza virus primes human polymorphonuclear leukocyte generation of superoxide. *Am.J.Respir.Cell.Mol.Biol.* 4, 347-354, 1991.

Calhoun, W.J., Dick, E.C., Schwartz, L.B. & Busse, W.W., A common cold virus, rhinovirus 16, potentiates airway inflammation after segmental antigen bronchoprovocation in allergic subjects. *J. Clin. Invest.* 94, 2200-8, 1994.

Chand, N., Harrison, J., Rooney, S., Pillar, J., Jakubicki, R., Nolan, K., Diamantis, W. & Sofia, R., Anti-IL-5 monoclonal antibody inhibits allergic late phase bronchial eosinophilia in guinea-pigs: a therapeutic approach. *Eur. J. Pharmacol.* 211, 121-123, 1992.

Choi, A.M.K. & Jacoby, D.B., Influenza virus A infection induces interleukin-8 gene expression in human airway epithelial cells. *FEBS Letters.* 309, 327-329, 1992.

Christensen, J.P., Anderson, E.C., Scheynius, A., Marker, O. & Thomsen, A.P., Alpha(4) integrin directs virus-activated CD8(+) T cells to sites of infection. *J. Immunol.* 154, 5293-5301, 1995.

Cockcroft, D., Ruffin, R. & Hargreave, F., Effect of SCH 1000 in allergen induced asthma. *Clin. Allergy* 8, 361-72, 1978.

Cockcroft, D.W., Ruffin, R.E. Dolovich, J. & Hargreave, F.E., Allergen-induced increases in non-allergic bronchial reactivity. *Clin. Allergy* 7, 505-513, 1977.

Costello, R.W., Schofield, B.H., Kephart, G.M., Gleich, G.J., Jacoby, D.B. & Fryer, A.D., Localization of eosinophils to airway nerves and effect on neuronal M2 muscarinic receptor function. *Am. J. Physiol.* 273, L93-103, 1997.

Coyle, A.J., Erard, F., Bertrand, C., Walti, S., Pircher, H. & Le, G.G., Virus-specific CD8+ cells can switch to interleukin-5 production and induce airway eosinophilia. *J. Exp. Med.* 181, 1229-33, 1995.

Douglass, J.A., Dhami, D., Gurr, C.E., Bulpitt, M., Shute, J.K., Howarth,

P.H., Lindley, I.J., Church, M.K. & Holgate, S.T., Influence of interleukin-8 challenge in the nasal mucosa in atopic and nonatopic subjects. *Am. J. Respir. Crit. Care Med.,* 150, 1108-13, 1994.

Dusser, D.J., Jacoby, D.B., Djokic, T.D., Rubenstein, I., Borson, D.B., & Nadel, J.A., Virus induces airway hyperresponsiveness to tachykinins: role of neutral endopeptidase. *J. Appl. Physiol.* 67, 1504-1511, 1989.

Einarsson, O, Geba, G.P., Zhu, Z., Landry, M. & Elias, J.A., Interleukin-11: stimulation *in vivo* and *in vitro* by respiratory viruses and induction of airways hyperesponsivenesss. *J. Clin. Invest.* 97, 915-924, 1996.

Eiser, N., & Guz, A., Effect of atropine on experimentally induced airway obstruction in man. *Bull. Europ. Physiopath. Resp.* 18, 449-60, 1982.

Elbon, C.L., Jacoby, D.B. & Fryer, A.D., Pretreatment with an antibody to interleukin-5 prevents loss of pulmonary M_2 muscarinic receptor function in antigen challenged guinea-pigs. *Am. J. Respir. Cell Mol. Biol.* 12, 320-328, 1995.

Empey, D.W., Laitinen, L.A., Jacobs, L., Gold, W.M., & Nadel, J.A., Mechanisms of bronchial hyperreactivity in normal subjects following upper respiratory tract infection. *Am. Rev. Respir. Dis.* 113, 523-527, 1976.

Evans, C.M., Jacoby, D.B., Gleich, G.J., Fryer, A.D., & Costello, R.W., Antibody to eosinophil major basic protein protects M_2 receptor function of antigen challenged guinea pigs *in vivo*. *J. Clin. Invest.* 100, 2254-2262, 1997.

Fish, J.E., Rosenthall, R.R., Summer, W.R., Menkes, H., Norman, P.S. & Permutt, S., The effect of atropine on acute antigen-mediated airway constriction in subjects with allergic asthma. *Am. Rev. Respir. Dis.* 115, 371-379, 1977.

Frick, O.L., German, D.F., & Mills, J., Development of allergy in children. I. Association with virus infections. *J. Allergy Clin. Immunol.* 63, 228-241, 1979.

Frigas, E., Loegering, D.A. & Gleich, G.J., Cytotoxic effect of the guinea pig major basic protein on tracheal epithelium. *Lab. Invest.* 42, 35-43, 1980.

Frigas, E., Loegering, D.A., Solley, G.O., Farrow, G.M. & Gleich, G.J, Elevated levels of the eosinophil granule MBP in the sputum of patients with bronchial asthma. *Mayo Clin. Proc.* 56, 345-353, 1981.

Fryer, A. D., El-Fakahany, E.E. & Jacoby, D.B., Parainfluenza virus type 1 reduces the affinity of agonists for muscarinic receptors in guinea-pig heart and lung. *Eur. J. Pharmacol.* 181, 51-58, 1990.

Fryer, A.D., Elbon, C.L., Kim, A.L., Xiao, H.Q., Levey, A.I. & Jacoby, D.B., Cultures of airway parasympathetic nerves express function M_2 muscarinic receptors. *Am. J. Respir. Cell. Mol. Biol.* 15, 716-725, 1996.

Fryer, A.D., & Jacoby, D.B., Function of pulmonary M_2 muscarinic receptors in antigen challenged guinea-pig is restored by heparin an poly-1-glutamate. *J. Clin. Invest.* 90, 2292-2298, 1992.

Fryer, A.D., & Jacoby, D.B., Parainfluenza virus infection damages inhibitory M_2 muscarinic receptors on pulmonary parasympathetic nerves in the guinea-pig. *Br. J. Pharmacol.* 102, 267-271, 1991.

Fryer, A.D. & Maclagan, J., Ipratropium bromide potentiates bronchoconstriction induced by vagal nerve stimulation in the guinea-pig. *Eur. J. Pharmacol.* 139, 187-191, 1987.

Fryer, A.D., Maclagan, J., Muscarinic inhibitory receptors in pulmonary parasympathetic nerves in the guinea-pig. *Br. J. Pharmacol.* 83, 973-978, 1984.

Fryer, A.D., & Wills-Karp, M., Dysfunction of M_2 muscarinic receptors in pulmonary parasympathetic nerves after antigen challenge in guinea-pigs. *J. Appl. Physiol.* 71, 2255-2261, 1991.

Gambone, L.M., Elbon, C.L., & Fryer, A.D., Ozone-induced loss of neuronal M_2 muscarinic receptor function is prevented by cyclophosphamide. *J. Appl. Physiol.* 77, 1492-1499, 1994.

Garofalo, R., Kimpen, J.L., Welliver, R.C., & Ogra, P.L., Eosinophil degranulation in the respiratory tract during naturally acquired respiratory syncytial virus infection. *J. Pediatr.* 120, 28-32, 1992.

Gies, J.-P., & Landry, Y., Sialic acid is selectively involved in the interaction of agonists with M_2 muscarinic acetylcholine receptors. *Biochem. Biophys. Res. Comm.* 150, 673-680, 1988.

Gulbenkian, A.R., Egan, R.W., Fernandez, X., Jones, H., Kreutner, W., Kung, T., Payvandi, F., Sullivan, L., Zurcher, J., & Watnick, A.S., Interleukin-5 modulates eosinophil accumulation in allergic guinea pig lung. *Am. Rev. Respir. Dis.* 146, 263-265, 1992.

Gundel, R.H., Lettes, L.G. & Gleigh, G.J., Human eosinophil major basic protein induces airway constriction and airway hyperresponsiveness in primates. *J. Clin. Invest.* 132, 599-606, 1991.

Haddad, E.B., Rousell, J., Mak, J.C. & Barnes, P.J., Transforming growth factor-beta 1 induces transcriptional down-regulation of M_2 muscarinic receptor gene expression. *Mol. Pharmacol.* 49, 781-7, 1996.

Haddad, E.B., Rousell, J., Mak, J.C.W., Adcock, I. & Barnes, P.J., Dexamethasone inhibits protein kinase C-induced down-regulation of M_2 muscarinic receptors in human lung fibroblasts. *Am. Rev. Respir. Crit. Care Med.* 151, A164, 1995.

Hamilton, S.E., McKinnon, L.A., Jackson, D.A., Goldman, P.S., Migeon, J.C., Habecker, B.A., Thomas, S.L., & Nathanson, N.M., Molecular analysis of the regulation of muscarinic receptor expression and function. *Life Sciences* 56, 939-943, 1995.

Henderson, F.W., Clyde, W.A., Collier, A.M., Denny, F.W., Senior,

R.J., Sheaffer, C.I., Conley, W.G., & Christian, R.M., The etiologic and epidemiologic spectrum of bronchiolitis in pediatric practice. *J. Pediatr.* 95, 183-190, 1979.

Holtzman, M., Mc Namara, P., Sheppard, D., Fabbri, L., Hahn, H., Graf, P. & Nadel, J., Intravenous versus inhaled atropine for inhibiting bronchoconstrictor responses in dogs. *J. Appl. Physiol.* 54, 134-39, 1983.

Hu, J., Wang, S.-Z., Forray, C., & El-Fakahany, E.E., Complex allosteric modulation of cardiac muscarinic receptors by protamine: a potential model for putative endogenous ligands. *Mol. Pharmacol.* 42, 311-324, 1992.

Hurd, J. & Heath, R.B., Effect of cyclophosphamide on infections in mice caused by virulent and avirulent strains of influenza virus. *Infect. Immun.* 11, 886-889, 1975.

Itkin, I. & Anand, S., The role of atropine as a mediator blocker of induced bronchial obstruction. *J. Allergy Clin. Immunol.* 45, 178-86, 1970.

Ito, Y. & Yoshitomi, T., Autoregulation of acetylcholine release from vagus nerves terminals through activation of muscarinic receptors in the dog trachea. *Br. J. Pharmacol.* 93, 636-646, 1988.

Jacoby, D.B., Gleich, G.J., & Fryer, A.D., Human eosinophil major basic protein is an endogenous allosteric antagonist at the inhibitory muscarinic M_2 receptor. *J. Clin. Invest.* 91, 1314-1318, 1993.

Jacoby, D.B., Gleich, G. J., & Fryer, A.D., Interaction of human eosinophil major basic protein with cloned human muscarinic receptors. *Am. J. Respir. Crit. Care. Med.* 151, A821, 1995.

Jacoby, D.B., Tamaoki, J., Borson, D.B. & Nadel, J.A., Influenza infection causes airway hyperresponsiveness by decreasing enkephalinase. *J. Appl. Physiol.* 64, 2653-2658, 1988.

Jacoby, D.B., Xiao, H.Q., Lee, N.H., Chan-Li, Y., Fryer, A.D., Virus- and interferon-induced loss of M_2 muscarinic receptor function and gene expression in cultured airway parasympathetic neurons. *J. Clin. Invest.* 102, 242, 1998.

Johnston, S.L., Pattemore, P.K., Sanderson, G., Smith, S., Lampe, F., Josephs, L., Symington, P., O'Toole, S., Myint, S.H., Tyrrell, D. & Holgate, S.T., Community study of role of viral infections in exacerbations of asthma in 9-11 year old children. *Br. Med. J.* 310, 1225-1229, 1995.

Killingsworth, C.R., Mingfu, Y. & Robinson, N.E., Evidence for the absence of a functional role for muscarinic M_2 inhibitory receptors in cat trachea *in vivo*; contrast with *in vitro* results. *Br. J. Pharmacol.* 105, 263-270, 1992.

Kresten, W., Protective effect of metered aerosol SCH 1000 (ipratropium bromide) against bronchoconstriction by allergen inhalation. *Respiration* 31, 412-17, 1974.

Lefort, J., Nahori, M.A., Ruffie, C., Vargaftig, B.B. & Pretolanti, M., *In vivo* neutralization of eosinophil-derived major basic protein inhibits antigen-induced bronchial hyperreactivity in sensitized guinea pigs. *J. Clin. Invest.* 97, 1117-1121, 1996.

Lemanske, R.F., Dick, E.C., Swensen, C.A., Vrtis, R. F., & Busse, W.W., Rhinovirus upper respiratory infection increases airway hyperreactivity and late asthmatic reactions. *J. Clin. Invest.* 83, 1-10, 1989.

Little, J.W., Hall, W.J., Douglas, R.G., Mudholkar, G.S., Speers, D.M. & Patel, K., Airway hyperreactivity and peripheral airway dysfunction in influenza A infection. *Am. Rev. Respir. Dis.* 118, 295-303, 1978.

McCaig, D.J., Comparison of autonomic responses in the trachea isolated from normal and albumin-sensitive guinea-pigs. *Br. J. Pharmacol.* 92, 809-816, 1987.

Minette, P. & Barnes, P.J., Prejunctional inhibitory muscarinic receptors on cholinergic nerves in human and guinea-pig airways. *J. Appl. Physiol.* 64, 2532-2537, 1988.

Minette, P.J., Lammers, J.W.J., Dixon, C.M.S., McCusker, M.T. & Barnes, P.J., A muscarinic agonist inhibits reflex bronchoconstriction in normal but not asthmatic subjects. *J. Appl. Physiol.* 67, 2461-65, 1989.

Noah, T.L. & Becker, S., Respiratory syncytial virus-induced cytokine production by a human bronchial epithelial cell line. *J. Physiol.* 265, L472-L478, 1993.

Okayama, M., Shen, T., Midorikawa, J., Lin, T.J., Inoue, H., Takishima, T. & Shirato, K. Effect of pilocarpine on propranolol-induced bronchoconstriction in asthma. *Am. J. Respir. Crit. Care Med.* 149, 76-80, 1994.

Openshaw, P.J., Clarke, S.L. & Record, F.M., Pulmonary eosinophilic response to respiratory syncytial virus infection in mice sensitized to the major surface glycoprotein G. *Int. Immuno.* 4, 493-500, 1992.

Orehek, J., Gayard, P., Grimaud, C. & Charpin, J., Allergic bronchoconstriction in asthmatics: The effect of a synthetic anticholingic agent. *Bull. Physiopathologie Respiratoire de Nancy* 11, 193, 1975.

Peterson, G.L., Rosenbaum, L.C., Broderick, D. J. & Schimerlik, M.I., Physical properties of the purified cardiac muscarinic acetylcholine receptor. *Biochemistry* 25, 3189-3202, 1986.

Piedimonte, G., Umeno, E., McDonald, D.M. & Nadel, J.A., Sendai virus infection potentiates neurogenic inflammation in the rat trachea. *J. Appl. Physiol.* 68, 754-760, 1990.

Rosenthal, R., Norman, P., Summer, W. & Permutt, S., Role of the parasympathetic system in antigen induced bronchospasm. *J. Appl. Physiol.* 42, 600-606, 1977.

Rosoff, M.L., Wei, J. & Nathanson, N.M., Isolation and characterization of the chicken M_2 acetylcholine receptor promoter region: induction of gene transcription by leukemia inhibitory factor and ciliary neurotrophic factor. *Proc. Natl. Acad. Sci. USA 93*, 14889-94, 1996.

Rousell, J., Haddad, E.B., Mak, J.C.W. & Barnes, P.J., Transcriptional down-regulation on M_2 muscarinic receptor gene expression in human embryonic lung (hel 299) cells by protein kinase c. *J. Biol. Chem.* 270, 7213-7218, 1995.

Ruffin, R., Cockroft, D. & Hargreave, F., A comparison of the protective effect of fenoterol and SCH 1000 on allergen induced asthma. *J. Allergy Clin. Immunol.* 61, 42-47, 1978.

Samuel, C.E., Antiviral actions of interferon: Interferon-regulated cellular proteins and their surprisingly selective antiviral activities. *Virology* 183, 1-11, 1991.

Scheid, A., Caliguiri, L.A., Compans, R.W. & Choppin, P.W., Isolation of paramyxovirus glycoproteins. Association of both hemagglutinating and neuraminidase activities with the larger SV5 glycoprotein. *Virology* 50, 640-652, 1972.

Scherle, P.A., Palladino, G. & Gerhard, W., Mice can recover from pulmonary influenza virus infection in the absence of class-I restricted cytotoxic T-cells. *J. Immuno.* 148, 212-217, 1992.

Schiller, I. & Lowell, F., The effect of drugs in modifying the response of asthmatic subjects to inhalation of pollen extracts as determined by vital capacity measurements. *Ann Allergy* 5, 584, 1947.

Schleimer, R.P., Sterbinsky, S.A., Kaiser, J., Bickel, C.A., Klunk, D., Tomioka, K., Newman, W., Lucinskas, F.W., Gimbrone, M.A., McIntyre, B.W. & Bochner, B.S., IL-4 induces adherence of human eosinophils & basophils but not neutrophils to endothelium. Association with expression of VCAM-1. *J. Immunol.* 148, 1086-1092, 1992.

Schoenborn, C.A. & Marano, M., Current estimates from the National Health Interview Survey: United States, 1988. DHHS Publication No. PHS 89-1501. *Vital Health Stat.* 10, 173, 1989.

Schultheis, A., Bassett, D., & Fryer, A., Ozone-induced airway hyperresponsiveness and loss of neuronal M_2 muscarinic receptor function. *J. Appl. Physiol.* 76, 1088-1097, 1994.

Schultze-Weringhaus, G., Anticholinergic versus β adrenergic therapy in allergic airways obstruction. *Respiration* 41, 239-47, 1981.

Schweizer, R.C., Welmers, B.A., Raaijmakers, J.A., Zanen, P., Lammers, J.W. & Koenderman, L., RANTES- and interleukin-8-induced responses in normal human eosinophils: effects of priming with interleukin-5. *Blood* 83, 3697-704, 1994.

Sheppard, D., Epstein, J., Holtzman, M. & Boushey, H., Effect of route of delivery of atropine delivery on bronchospams from cold air and methacholine. *J. Appl. Physiol.* 54, 130-33, 1983.

Sigurs, N., Bjarnason, R. & Sigurbergsson, F., Eosinophil cationic protein in nasal secretion and in serum and myeloperoxidase in serum in respiratory syncytial virus bronchiolitis: relation to asthma and atopy. *Acta. Paediatr.* 83, 1151-5, 1994.

Singer, S.H., Ford, M. & Kirschstein, R.L., Respiratory diseases in cyclophophamide-treated mice. II. Decreased virulence of PR8 influenza virus. *Infect. Immun.* 5, 957-960, 1972.

Smolinski, P.A., Offermann, M.K., Eckman, J.R. & Wick, T.M., Double-stranded RNA induces sickle erythrocyte adherence to endothelium: a potential role for viral infection in vaso-occlusive pain episodes in sickle cell anemia. *Blood* 85, 2945-2950, 1995.

Subsauste, M.C., Jacoby, D.B. & Proud, D., Infection of a human respiratory epithelial cell line with rhinovirus. Induction of cytokine release and modulation of susceptibility to infection by cytkine exposure. *J. Clin. Invest.* 96, 549-557, 1995.

Viac, J., Schmitt, D. & Claudy, A., Expression of adhesion receptors in epidermal tumors - correlation with tnf-alpha expressing cells. *Anticancer Research* 15, 551-555, 1995.

Walsh, J.J., Dietlein, L.F., Low, F.N., Burch, G.E. & Mogabgab, W.J., Bronchotracheal response in human influenza. *Arch. Int. Med.* 108, 376-388, 1960.

Welliver, R.C., Upper respiratory infections in asthma. *J. Allergy Clin. Immunol.* 72, 341-346, 1983.

Wood, I.C., Roopra, A., Harrington, C. & Buckley, N.J., Structure of the M_4 cholinergic muscarinic receptor gene and its promoter. *J. Biol. Chem.* 270, 30933-30940, 1995.

Yu, D.Y.C., Galant, S.P. & Gold, W.M., Inhibition of antigen-induced bronchoconstriction by atropine in asthmatic patients. *J. Appl. Physiol.* 32, 823, 1972.

Chapter 5

ROLE OF CD4+ T CELLS IN THE DEVELOPMENT OF AEROALLERGEN-INDUCED ALLERGIC INFLAMMATION

Marsha Wills-Karp, Ph.D.

I. INTRODUCTION

Asthma is a chronic, often debilitating disease which currently afflicts 10-15 million people in the US alone. Alarming increases in the incidence, morbidity and mortality of this disease worldwide have been noted over the last decade. Asthma is characterized by airway hyperresponsiveness, eosinophilic inflammation and elevated IgE levels. Most asthma in childhood is allergic in nature [Cockcroft *et al* 1977; Sears *et al* 1991]. Longitudinal epidemiological studies of childhood asthmatics have demonstrated that sensitivity to common household allergens is significantly associated with the development of asthma [Sears *et al* 1989]. Inhalation of indoor allergens can exacerbate allergic asthma, and prolonged avoidance of exposure to allergens such as house dust mites and other indoor allergens (animal dander, cockroach allergens, and mold spores) leads to reductions in both airway inflammation and airway hyperresponsiveness in allergic asthmatics [Platts-Mills *et al* 1982]. Inhalation of indoor allergens may also predispose to the development of allergic asthma. Predisposition to asthma, and presumably to chronic inflammatory responses in the airways, has been linked closely to atopic status, which has a hereditable component [Hargreave *et al* 1986; Cockcroft *et al* 1984; Gereard *et al* 1978]. However, the asthmatic diathesis has also been linked to early childhood exposure to aeroallergens [Luoma *et al* 1983]. The factors which predispose certain individuals to mounting deleterious responses to otherwise harmless inhaled allergens are not well understood.

It has recently become appreciated that a subset of CD4+ T cells which has been distinguished functionally by its pattern of cytokine secretion, plays a key role in the pathogenesis of allergic asthma. This review will explore the role of this specific subset of T helper cells in the pathogenesis of asthma and in addition will explore possible mechanisms for aberrant expansion of this pathogenic subset of CD4+ T helper cells in response to common indoor allergens in allergic individuals.

II. STRUCTURAL AND FUNCTIONAL CORRELATES OF THE ALLERGIC PULMONARY RESPONSE TO INHALED ALLERGENS

Asthma is characterized structurally by the presence of chronic inflammation of the airways, with intense infiltration of the bronchial

1-56670-373-5/99/$0.00+$.50

mucosae by lymphocytes, eosinophils, and mast cells, along with epithelial desquamation [McFadden and Gilbert 1992; Beasley *et al* 1989]. Functionally, asthma is characterized by abnormal bronchoconstriction, with hyperresponsiveness of the airways in response to challenge to a variety of stimuli. These stimuli include cold air, exercise, allergen challenge, and diverse pharmacological agents [McFadden and Gilbert 1992; Pauwels 1989]. Structure and function interact at the level of airway geometry in that bronchoconstriction takes place in airways already narrowed by inflammation.

The degree of eosinophilia in both the blood and BAL fluid of asthmatics (both allergic and nonallergic) correlates with the clinical severity of disease as well as with the degree of baseline bronchial hyperresponsiveness [Bousquet *et al* 1990]. Although the actual mechanism(s) by which eosinophils exert pathophysiological effects on airways remains unclear, eosinophils are known to secrete a number of lipid mediators and basic proteins which have profound effects on a range of airway cell types [Gleich 1990; Wardlaw *et al* 1988; Gundel *et al* 1991]. Release of these mediators can cause increases in vascular permeability, bronchoconstriction, and the destruction of airway epithelial cells, all characteristic features of asthma [Gleich 1990; Wardlaw *et al* 1988; Gundel *et al* 1991].

In allergic asthma, airway challenge with antigen leads to the rapid onset of mucosal edema, increases in airway smooth muscle tone, airway narrowing, and an increase in neutrophil, mast cell, and eosinophil numbers in bronchoalveolar lavage (BAL) fluids [Pauwels 1989]. This immediate response is thought to be the result of the action of mediators such as histamine released by mast cells and/or basophils. Some allergic asthmatics also develop late phase responses beginning three to six hours after antigen challenge and lasting up to 48 hours in the absence of therapy. In these responses, airway narrowing is associated with BAL fluid which is characterized primarily by increased numbers of eosinophils [Pauwels 1989; Bousquet *et al* 1990; Cockcroft 1989; De Monchy *et al* 1985].

Allergic asthma is further characterized by the production of abnormal amounts of IgE in response to the inhalation of aeroallergens [Sears *et al* 1991; Sears *et al* 1989; Stenius *et al* 1971]. Serum levels of IgE correlate with the severity of disease in these patients. On the other hand, non-allergic asthmatics do not have elevated levels of serum IgE. As the basic pathophysiology of non-allergic asthma is otherwise similar to that of allergic asthma, the exact role of IgE in the development of airway hyperresponsiveness and inflammation is unclear.

III. THE ROLE OF T LYMPHOCYTES IN ALLERGEN- INDUCED AIRWAY INFLAMMATION & AIRWAY HYPERRESPONSIVENESS

As a primary regulator of the inflammatory cascade, the T lymphocyte has been implicated in the pathogenesis of allergic airway disease [Gerblich *et al* 1991; Azzawi *et al* 1990; Corrigan and Kay 1990]. Differences in pulmonary lymphocyte populations have been demonstrated, with elevated numbers of lymphocytes in bronchial mucosal biopsies from allergic asthmatics [Gerblich *et al* 1991; Azzawi *et al* 1990; Corrigan and Kay 1990]. Cell recruitment to the airways is suggested by the finding of increased numbers of CD4+ T cells in bronchoalveolar lavage (BAL) with concurrent decreased numbers of CD4+ T cells in peripheral blood following allergen challenge [Lundgren *et al* 1988]. Experimental data support a generalized increase in T cell activation in asthmatics, with increased expression of the interleukin-2 receptor (IL-2R), class II histocompatibility antigens (HLA-DR), and very late activation antigen-1 (VLA-1) in blood, BAL and bronchial biopsies [Corrigan and Kay 1990; Walker *et al* 1992].

Functional subsets of CD4+ T cells have been distinguished at both clonal and population levels by the unique profiles of cytokines that they produce [Mosmann *et al* 1986; Street and Mosmann 1991]. The differential presence of these cytokine phenotypes in a variety of allergic and infectious diseases of both mice and humans has provided both descriptive power and theoretical insight into disease pathogenesis [Peltz 1991; Kapsenberg *et al* 1991; de Prete *et al* 1991; Sher *et al* 1991; Clerici *et al* 1993; Yamamura *et al* 1991]. Th1 cells produce interleukin-2 (IL-2), TNF-ß, and interferon gamma (IFN-γ), but not interleukin-4, IL-5, IL-6, IL-9, or IL-10, and are critical in the development of cell-mediated immunity [Mosmann *et al* 1986; Street and Mosmann 1991]. Th2 cells produce the opposite set of cytokines and are important in the stimulation of IgE production, mucosal mastocytosis, and eosinophilia [Mosmann *et al* 1986; Street and Mosmann 1991]. Th1 and Th2 cells differ in their primary autocrine growth factors (IL-2 for Th1 cells, IL-4 for Th2 cells), as well as in their response to multiple pharmacologic agents and costimulatory signals from antigen presenting cells. The subsets show cross-regulation, with IFN-γ from Th1 cells down-regulating cytokine production and proliferation by Th2 cells, and IL-4 and IL-10 from Th2 cells providing a similar function for Th1 cells [Street and Mosmann 1991].

A possible immunopathogenic role for Th2 cells in asthma has been based on the role that these cytokines play in IgE synthesis and eosinophil regulation (Fig. 1). In particular, IL-4 may be of critical importance in allergic inflammatory processes, as it is required for the differentiation and expansion of Th2 lymphocytes and inhibition of the development of Th1 lymphocytes. In addition, IL-4 modulates the growth and differentiation of mast cells and regulates IgE synthesis [Madden *et al* 1991; Katona *et al* 1993; Finkelman *et al* 1988]. In turn, interleukin-5 controls the differentiation [Campbell *et al* 1987], recruitment [Wang *et al* 1989], and activation of eosinophils [Lopez *et al* 1988] (Fig. 1).

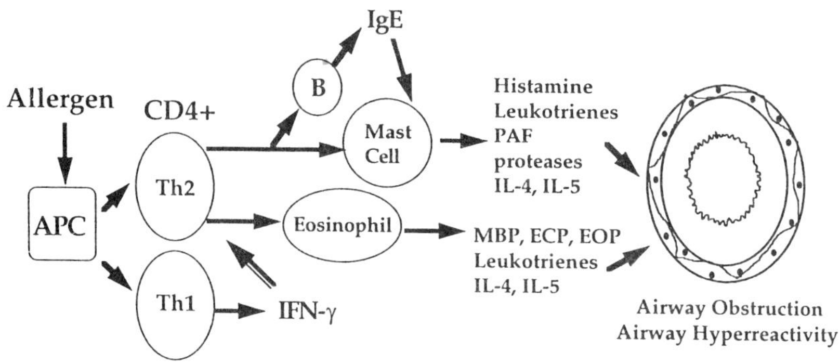

Figure 1. Hypothetical scheme of the role of CD4+ T cells in the pathogenesis of allergic asthma. Allergen presentation to CD4+ T lymphocytes in the context of MHCII molecules on antigen presenting cells (APC) results in differentiation of CD4+ T cells into one of two subsets defined by the profile of cytokines they secrete. Th1 cells secrete interferon (IFN-γ) and are important in cell-mediated immunity, whereas the Th2 subset secretes IL-4 and IL-5. Substantial evidence suggests that CD4+ T cells in the asthmatic lung are predominantly of the Th2 phenotype. This subset is thought to induce the pathology associated with allergic asthma through the actions of IL-4 and IL-5 on effector cells. IL-4 stimulates production of immunoglobulin E (IgE), which sensitizes mast cells and eosinophils. In turn, IL-5 is critical in mobilization of eosinophils from the bone marrow. Reexposure of sensitized effector cells to allergen results in the release of a number of mediators which act directly on target tissues within the lung, causing smooth muscle constriction, vascular leakage, and overproduction of mucus by the airway epithelium, all of which may contribute to the airway obstruction associated with the disease. Amplification and propagation of the type-2 immune response may occur as a result of the ability of these effector cells to produce additional IL-4 and IL-5. MBP-major basic protein, ECP-eosinophil cationic protein, EOP- eosinophil peroxidase, PAF-platelet activating factor, solid arrows represent stimulatory actions, striped arrows represent inhibitory actions.

Experimental support for the involvement of these cytokines in the pathogenesis of asthma is provided by the fact that unique cytokine patterns are expressed in T cells from asthmatic patients [Walker *et al* 1992; Robinson *et al* 1992]. T cells from both the BAL fluids and bronchial biopsies of allergic asthmatics express elevated levels of mRNA for IL-4 and IL-5 [Walker *et al* 1992; Robinson *et al* 1992]. Interestingly, successful steroid treatment has been shown to increase IFN-γ levels in the BAL of asthmatic patients, while decreasing IL-4 and IL-5 levels [Wilkinson *et al*

1991]. A longitudinal study of children has shown that the propensity to develop asthma was associated with low stimulated levels of IFN-γ at 9 months of age, suggesting that a type-1 response was protective [Martinez *et al* 1995]. Additional support for this hypothesis in humans has been provided in a population based study described by Shirakawa and colleagues [Shirakawa *et al* 1997]. They found that among Japanese schoolchildren there was a strong inverse association between delayed type hypersensitivity to *Bacillus Calmette Guerin* (BCG) and atopy. Positive tuberculin responses predicted a lower incidence of asthma, lower serum IgE levels, and cytokine profiles biased toward a type-1 profile. These results suggested that exposure and response to BCG may, by modification of immune cytokine profiles, inhibit atopic disorders.

Although considerable descriptive evidence suggests that CD4+ T-lymphocytes and Th2 cytokines are important in the pathogenesis of asthma, definitive proof is difficult to obtain in humans. Therefore, experimental animal models have been extremely useful in delineation of the role of both CD4+ T cells and T cell-derived cytokines in the pathogenesis of asthma. Direct evidence for a causal role for CD4+ T cells in the development of airway hyperresponsiveness was provided by my laboratory in a murine model of allergen-induced airway hyperresponsiveness [Gavett *et al* 1994]. In this model, ovalbumin sensitization and local lung challenge results in increases in airway responsiveness to acetylcholine challenge and an eosinophilic inflammatory response in the lung. Depletion of CD4+ T cells in sensitized mice prior to local lung antigen challenge with specific monoclonal antibodies prevented the development of allergen-induced airway allergic responses. These findings have been confirmed by several investigators [Nakajima *et al* 1992; Olivenstein *et al* 1994]. Likewise several lines of evidence suggest a pivotal role for Th2 cytokines in asthma. Firstly, studies in which IL-4 and IL-5 have been manipulated either through antibody blockade [Gavett *et al* 1997; Coyle *et al* 1995] or gene targeting [Brusselle *et al* 1995; Foster *et al* 1996] have shown that they are essential to the development of allergic inflammation and airway hyperresponsiveness. Secondly, we have shown that administration of agents such as IL-12 and IFN-γ which inhibit Th2 cytokine production and stimulate Th1 pathways, prevent the development of antigen-induced airway hyperresponsiveness [Gavett *et al* 1995]. Below we explore the possible mechanisms by which the type-2 cytokines, IL-4 and IL-5, play a role in the inflammatory response observed in asthma.

IV. MECHANISMS OF TH2-INDUCED AIRWAY ALLERGIC RESPONSES

The mechanisms by which Th2 cytokines mediate allergic inflammation and airway hyperresponsiveness are not currently well understood. Th2 cytokines may potentially induce airway hyperresponsiveness through direct or indirect effects on cells such as B cells, mast cells, and eosinophils. To my knowledge, no one has examined the direct effects of either IL-4 or IL-5 on airway reactivity in in vitro systems which are relatively devoid of cellular, neural and humoral components. Below we examine the potential mechanisms by which IL-4 and IL-5 may induce airway inflammation and airway hyperresponsiveness, independently.

A. INTERLEUKIN-4-INDUCED ALLERGIC INFLAMMATION

IL-4, is a 18-20 kd glycoprotein, produced by T cells, mast cells and basophils [Bradding *et al* 1992; Seder *et al* 1991]. As discussed above, IL-4 has been shown to be essential for Th2 cell expansion and growth [Swain *et al* 1990; Hsieh *et al* 1992]. In addition to its pivotal role in differentiation of Th2 cells, IL-4 also induces a plethora of cellular responses which are potentially important in the development of allergic airway diseases including: its ability to regulate the synthesis of IgE [Finkelman *et al* 1988]; its role in stimulation of the growth and development of mast cells, in conjunction with IL-3 [Madden *et al* 1991; Katona *et al* 1993]; and its role in induction of VCAM-1 expression on the vascular endothelium [Schleimer *et al* 1992]. The cellular actions of IL-4 are mediated via a cell surface receptor composed of a cytokine-specific α chain and the common γc chain used by several other cytokines [Lin *et al* 1995]. Ligation of the IL-4 receptor results in the activation of at least two distinct signalling pathways [Witthuhn *et al* 1994; Keegan *et al* 1994; Sun *et al* 1995]. One involves the activation of signal transducer and activator of transcription factor 6 (Stat6) through phosphorylation by JAK1 and JAK3 [Kaplan *et al* 1996]. Once activated, Stat6 proteins form homodimers, translocate to the nucleus, and directly bind to specific promoter sequences to regulate gene transcription. In addition to Stat6 activation, stimulation of the IL-4 receptor has also been shown to induce the phosphorylation of an insulin receptor substrate (IRS) termed IRS-2 [Sun *et al* 1995]. Activated IRS-2 is thought to be responsible for T cell proliferation in response to IL-4.

Due to IL-4's plethora of actions it has been difficult to ascertain the specific mechanisms by which it may mediate the pathology of allergic disease. Specifically, since IL-4 is required early during the immune response to direct T cell differentiation into the type-2 pattern, it is difficult to separate this event from any effects that IL-4 may exert on effector cells such as B cells, mast cells and eosinophils. However, my colleagues and I [Gavett *et al* 1997] have shown that blockade of the IL-4 receptor at the time of local lung challenge was effective at suppressing airway hyperresponsiveness and eosinophilic inflammation, despite continued elevations in Th2 cytokine production. These results suggested that in addition to being necessary for Th2 differentiation, IL-4 must also work through activation of receptors on cells other than T cells such as B cells, mast cells, eosinophils and perhaps airway epithelial cells during specific immune responses to inhaled antigens.

One of the potential effector mechanism(s) by which IL-4 may induce allergic inflammation and airway function changes is through its role as the primary inducer of immunoglobulin class switching in B cells leading to the synthesis and secretion of IgE [Finkelman *et al* 1988]. One of the defining features of allergic asthma is consistently elevated levels of serum IgE [Sears *et al* 1991]. IgE is thought to play a role in allergic inflammation by preferentially binding to high-affinity Fc-epsilon receptors on mast cells and basophils, thereby sensitizing them. When Fc-epsilon receptors are cross-linked by subsequent IgE binding to the allergen, histamine and other pharmacological mediators responsible for immediate phase responses are released from the sensitized cells. Activated mast cells and basophils have recently been shown to release a number of cytokines which are important in late phase allergic response such as, IL-4 and IL-5 [Bradding *et al* 1992; Seder *et al* 1991]. Thus, IgE and mast cells may play an important role in initiation and propagation of airway inflammation. On the other hand, studies with mice devoid of B cells [Korsgren *et al* 1996] or IgE [Melhop *et al* 1997] have suggested that neither IgE nor mast cells are essential for the development of eosinophilia inflammation or airway hyperresponsiveness. Further studies are needed to more clearly define the role of IgE in the asthmatic diathesis.

Eosinophilic inflammation is clearly a hallmark of allergic asthma and considerable evidence suggests an association between pulmonary eosinophil infiltration and airway hyperresponsiveness in human asthma [Wardlaw *et al* 1988]. Eosinophils are postulated to induce airway hyperresponsiveness through release of cationic proteins, eosinophil cationic protein (ECP), and major basic protein, which are thought to damage the airway epithelium leaving smooth muscle more susceptible to contractile mechanisms [Frigas *et al* 1980; Gleich 1990]. IL-4 has been shown to be important in the recruitment of eosinophils into sites of local inflammation [Gavett *et al* 1997; Brusselle *et al* 1995; Rankin *et al* 1996]. The role of IL-4 in pulmonary eosinophilia has been demonstrated in IL-4 transgenic mice which display spontaneous airway eosinophilia [Rankin *et al* 1996]. Conversely, IL-4 deficient animals do not develop pulmonary eosinophilia following aerosol antigen challenge [Brusselle *et al* 1995]. Although the exact mechanism(s) of IL-4's ability to induce tissue eosinophilia are unknown, several potential mechanisms exist. IL-4 may induce eosinophilia by: 1) promoting the production of IL-5; 2) upregulating vascular cell adhesion molecule-1 (VCAM-1) expression on vascular endothelium [Schleimer *et al* 1992]; or 3) stimulating chemokine production [Mochizuki *et al* 1998]. IL-4 is clearly important in stimulating Th2 cytokine production and thereby may regulate tissue eosinophilia indirectly through its ability to enhance IL-5 production. This issue will be addressed in more detail below. In addition, IL-4 is thought to upregulate VCAM-1 expression which upon interaction with the very late activation antigen-4 (VLA-4) on endothelial cells promotes eosinophil attachment and migration. Although several animal studies have supported a role for VCAM-1 expression in tissue eosinophilia [Nakajima *et al* 1994; Pretolani *et al* 1994], it is not clear whether IL-4 is the primary stimulus of its expression. For example, Nakajima et al [Nakajima *et al* 1994] showed that blockade of VCAM-1 and VLA-4 inhibited antigen-induced eosinophil infiltration into the mouse trachea, but that anti-IL-4 antibody administration only partially inhibited VCAM-1 expression. Furthermore, my group [Kuperman *et al* 1998] has recently shown that allergen-induced VCAM-1 expression is independent of the IL-4-activated Stat6 signalling pathway. An alternative mechanism for IL-4's apparent role in eosinophil recruitment is suggested by a recent report by Mochizuki and colleagues [Mochizuki *et al* 1998] which demonstrates that IL-4 stimulates fibroblast production of the chemokine, eotaxin.

Excessive production of airway mucus glycoproteins is a consistent finding in the lungs of asthmatics, in particular in those patients who die in *status asthmatics* [Aikawa *et al* 1992]. Evidence for a role for IL-4 in mucus production has been provided by IL-4 transgenic mice who exhibit profound increases in mucus containing cells in the airway epithelium [Rankin *et al* 1996]. This finding is supported by a report by my laboratory [Gavett *et al* 1997] demonstrating that antigen-induced increases in mucus containing cells was ablated by blockade of the IL-4 receptor. Furthermore, we [Kuperman *et al* 1998] have recently shown that the airways of Stat6 deficient animals do not show evidence of mucus production following allergen provocation. Whether the effect of IL-4 on mucus production is direct or indirect is currently unknown. Savoie *et al* 1995 had shown that mucus secretion after antigen challenge in guinea pigs correlated with eosinophil peroxidase levels and with the number of eosinophils in the BAL fluid. However, in the study of Stat6 deficient mice, no mucus containing cells were present in the lungs of allergen-exposed Stat6-deficient animals, despite the presence of eosinophils, suggesting that differentiation of airway epithelial cells to mucus-containing cells is likely a result of direct IL-4 stimulation [Pretolani *et al* 1994]. This hypothesis is consistent with the demonstration that IL-4 induces expression of a specific mucin gene, MUC5AC in the airways of IL-4 transgenic mice [Temann *et al* 1997]. Further studies will be required to determine the exact mechanisms of IL-4-stimulated mucus production and its contribution to functional changes observed in asthma.

Taken together, these studies demonstrate the importance of IL-4 in mediating inflammation and constriction of the airways in response to antigen exposure. Due to the pleiotropic effects of IL-4 on specific immune responses to antigen exposure, the exact mechanisms of its action remain to be defined. However, these studies do suggest the possibility that genes important in the regulation, production of, or responsiveness to IL-4, may underlie allergic asthma. In support of this hypothesis, it has been shown that the inheritance of atopic asthma has been linked to the gene cluster on chromosome 5q containing the genes for IL-4 and IL-5 [Marsh *et al* 1994; Postma *et al* 1995]. Furthermore, Hershey *et al* 1997 have recently demonstrated that expression of a mutant form of the IL-4 receptor α chain in allergic patients is associated with increased IL-4 signalling. The association of the novel IL-4 receptor allele with atopic disease will likely prompt studies of other candidate genes in the IL-4 signalling pathway such as the Stat6 transcription factor.

B. ROLE OF IL-5 IN ALLERGEN-INDUCED AIRWAY HYPERRESPONSIVENESS

Interleukin-5 is likely a pivotal cytokine in the development of allergic asthma, because it has been shown to be the primary determinant of eosinophil differentiation, activation and survival [Campbell *et al* 1987; Wang *et al* 1989; Lopez *et al* 1988]. In bone marrow, IL-5 is important for stimulation of eososinophilopoiesis and promoting the terminal differentiation of myeloid precursors into eosinophils. IL-5 also increases eosinophil adhesion to vascular endothelial cells, promotes the migration of eosinophils from the blood into the tissue, prolongs eosinophil survival in tissues and augments the cytotoxic activity of eosinophils. Furthermore, IL-5 may activate pulmonary eosinophils and cause the release of their cytotoxic products [Lopez *et al* 1988]. As stated above, IL-5 is derived from CD4+ T cells, but more recently has been shown to be released by mast cells and eosinophils themselves [Weller *et al* 1996]. The importance of IL-5 in antigen-induced eosinophilia has been well documented in several species [Van Oosterhout *et al* 1993; Kung *et al* 1995]. For example, blockade of endogenous IL-5 levels in antigen-sensitized guinea pigs [van Oosterhout *et al* 1993] and mice [Kung *et al* 1995] has resulted in significant suppression of both BAL and tissue eosinophilia. Consistent with these observations, intratracheal administration of IL-5 has been shown to induce eosinophil accumulation in the guinea pig lung in vivo [van Oosterhout *et al* 1993]. More definitely, mice in which the IL-5 gene has been disrupted do not develop aeroallergen-induced pulmonary eosinophilia [Foster *et al* 1996]. Reconstitution of these mice with IL-5 by administration of recombinant vaccinia viruses engineered to express IL-5, completely restored aero-allergen induced eosinophilia. In addition to IL-5, other mediators such as eotaxin, MCP-5 and RANTES are also important for eosinophil migration in tissues. These observations may appear to be in conflict with the results of the above studies; however, it is thought that IL-5 is essential for stimulating the release of a pool of bone marrow eosinophils and that recruitment of eosinophils to the site of inflammation may be dependent on chemokines such as eotaxin. Once the cells have accumulated in the tissues, locally produced IL-5 may be of further importance in enhancing eosinophil survival and responsiveness.

Although the relationship between IL-5 and eosinophilia is well-established, the relationship between eosinophilia and airway hyperresponsiveness is not as clear. For example, Corry and colleagues [Corry *et al* 1996] demonstrated that anti-IL-5 antibody administration to ovalbumin-sensitized mice at levels that suppressed eosinophils had no effect on subsequent antigen-induced airway reactivity. Furthermore, Van Oosterhout *et al* 1993 demonstrated that administration of IL-5-producing cells into the peritoneal cavity of guinea pigs produced airway hyperresponsiveness several days before any evidence of eosinophil airway recruitment. Lastly, IL-5 transgenic mice exposed to ovalbumin show a marked eosinophilia, but do not develop airway hyperresponsiveness [Lefort *et al* 1996]. On the other hand, studies of IL-5 deficient mice support a role for IL-5 and eosinophils in the development of allergen-induced airway hyperresponsiveness [Foster *et al* 1996]. A potential explanation for these discrepancies may be provided by Kraneveld and coworkers [Kraneveld *et al* 1997] who demonstrate that IL-5 is indeed important in airway hyperresponsiveness, but perhaps not via recruitment of eosinophilia alone but through release of neuropeptides capable of inducing bronchoconstriction and airway hyperresponsiveness to contractile agents.

Collectively, these studies support a role for Th2 cytokines in the pathogenesis of allergic asthma. Further studies are needed to determine the exact mechanisms by which T cell-derived cytokines induce alterations in the regulation of allergic inflammation and airway hyperresponsiveness, as well as the mechanisms that drive the differentiation of Th2 cells in response to aero-allergen exposure in susceptible individuals.

V. REGULATION OF CD4+ T CELL DIFFERENTIATION

Although considerable evidence supports a pathogenic role for Th2 cytokines in asthma, very little is known about the underlying cause(s) of the aberrant expansion of Th2 cells. To date, the ontogeny of Th1 and Th2 subsets remains unclear. The data from studies using T cell receptor transgenic mice [Seder *et al* 1992; Hsieh *et al* 1993], suggest that naive T helper cells have the ability to differentiate into either Th1 or Th2 cells. Naive T cells generate IL-2 when activated by antigen presentation and costimulatory signals (i.e. B7 molecules, CD58, and CD40), and progress through a multipotential cell that generates a mixed spectrum of cytokines, including IL-2, IL-4, and IFN-γ [Locksley 1994]. Although, many immune responses likely retain the (ThO) phenotype, responses may be polarized to either the Th1 or Th2 type, a process which is markedly influenced by the nature of the antigen, the dose of antigen, and the cytokine milieu during initial priming by specific antigens [Locksley 1994]. The elegant use of mice transgenic for a specific alpha/beta T cell receptor has provided details of this critical involvement of the cytokine environment on subset differentiation during naive lymphocyte priming [Seder *et al* 1992; Hsieh *et al* 1993]. In these models, IL-12 directly primes CD4+ T cells for Th1 differentiation [Seder *et al* 1992]. Th2 differentiation, in turn, is critically dependent upon IL-4 [Hsieh *et al* 1993].

A. INTERLEUKIN-4'S ROLE IN TH2 DIFFERENTIATION

Interleukin-4 as discussed above, is essential for the differentiation of CD4+ T cells into Th2 cells and is required for the expansion of Th2 cell clones, while inhibiting Th1 cell expansion. Although considerable evidence supports the critical role of IL-4 in differentiation of T cells towards a Th2 phenotype, little is known about the initial source of IL-4 production which is needed to drive the Th2 response. In addition to T cells, other cell types have been shown to produce IL-4 such as mast cells [Bradding *et al* 1992; Seder *et al* 1991], basophils [Seder *et al* 1991], and a CD4+ T cell that expresses the phenotype CD4+ $NK1.1^+$ [Yoshimoto and Paul 1994]. It has been hypothesized that CD4+ T cells expressing the $NK1.1^+$ marker might be the source of early IL-4 required for Th2 cell differentiation. These cells have been shown to rapidly produce IL-4 after in vivo treatment with anti-CD3 antibodies in contrast to conventional alpha-beta T cells [Yoshimoto *et al* 1994]. Several in vivo studies support a role for these $NK1.1^+$ T cells in providing the initial source of IL-4 to naive T cells [Yoshimoto *et al* 1994; Yoshimoto *et al* 1995], while other studies have not supported this conclusion [Brown *et al* 1996]. For example, Brown and coworkers [Brown *et al* 1996] examined the immune response of ß2-macroglobulin deficient animals, which don't express NK1.1+ T cells, to a variety of antigens which have been shown to induce potent type-2 immune responses. They found that production of IL-4 and the generation of IgE and eosinophil responses were preserved as assessed by a variety of assays. These studies would suggest that beta 2-m dependent NK1.1+ CD4+ T cells may not be required for Th2 development in vivo. On the other hand, studies using passive transfer of CD4+ T cells into mice in which the IL-4 gene had been disrupted suggested that CD4+ T cells alone were sufficient for the generation of type-2 immune responses (i.e. IgE production) upon subsequent antigen challenge [Schmitz *et al* 1994]. These studies suggested that neither mast cells or basophils provided the initial IL-4 necessary for Th2 cell differentiation. Perhaps mast cell and basophil production of IL-4 serves to amplify the response later during immune responses.

Other studies suggest that the initial source of IL-4 is the CD4+ T cell itself, but that it is not IL-4 dependent. A recent study of Leishmanial infection has shown that Balb/c mice which develop type-2 cytokine responses to infection produce IL-4 early during the infection and that levels of IL-4 then return to baseline and are elevated again later during infection

[Launois *et al* 1997]. Anti-IL-4 antibody treatment prior to infection does not ablate the initial production of IL-4, thus suggesting that the initial production of IL-4 may not be controlled by IL-4 itself as it is in the subsequent elevation. These results suggest that the initial IL-4 may be CD4+ T cell derived, but may be initiated by other signals.

A recent study has shown that IL-6, derived from antigen presenting cells, initiates the polarization of naive CD4+ T cells to effector Th2 cells by inducing the production of endogenous IL-4 [Rincon *et al* 1997]. In this study they demonstrated that IL-6 treatment of T cells resulted in upregulation of IL-4 gene expression. They speculate that IL-6 does so by activating a family of transcription factors called NF-IL-6, which have been shown to have three functional binding sites in the IL-4 promoter [Davydov *et al* 1995]. In support of this hypothesis elevated levels of IL-6 have been found in the BALF of allergic asthmatics [Mattoli *et al* 1991]. In addition to its potential role in IL-4 production, IL-6 has a number of proinflammatory actions including its ability to stimulate proliferation of thymocytes and mature T cells , upregulate IL-4 dependent IgE production, and mediate the differentiation of B cells [Van Snick *et al* 1990]. Interestingly, IL-6 transgenic mice have been shown to spontaneously develop a mononuclear cell infiltrate in the lung consisting predominantly of CD4+ T cells and B cells [DiCismo *et al* 1995]. The response to allergen challenge in these transgenic animals has not been examined. Elucidation of the initial polarizing source of IL-4 or the signals required for its production early during an immune response remains an important goal for understanding and determining the factors important in susceptibility to the development of pathogenic type-2 immune responses to aero-allergen exposure.

B. INTERLEUKIN-12 AS THE PRIMARY DETERMINANT OF TH1 CELL DIFFERENTIATION

Interleukin-12, a product of macrophages and other antigen presenting cells, is required for expansion of Th1 clones [Trinchieri *et al* 1994]. IL-12 is a 75kD cytokine composed of the covalently linked, glycosylated products of two separate genes: a heavy chain unit (p40) whose expression is tightly regulated in cells which secrete the intact heterodimer and a light chain subunit (p35) which is constitutively expressed at low levels by most cell types that have been studied [Trinchieri *et al* 1994]. Interestingly, the gene encoding the p40 subunit is located in the region of chromosome 5 in which linkage to human asthma has been demonstrated [Postma *et al* 1995]. IL-12 has been shown to inhibit production of IL-4 and IL-10 in allergen-specific human CD4+ T lymphocytes [Romagnani 1994]. In a murine model of asthma, we have [Gavett *et al* 1995] demonstrated that exogenous administration of recombinant IL-12 to allergen-sensitized animals both prevented and reversed the development of allergic airway responses. Protection from allergen-induced alterations in airway function was associated with ablation of type-2 cytokine production and an enhancement of IFN-γ levels. Coadministration of rIL-12 and anti-IFN-γ antibodies indicated that IL-12's inhibitory effects on airway hyperresponsiveness were mediated via stimulation of IFN-γ (Fig. 2). Other investigators have subsequently confirmed these studies and have also demonstrated that IL-12 suppresses antigen-specific increases in serum IgE levels in murine models of asthma [Kips *et al* 1996].

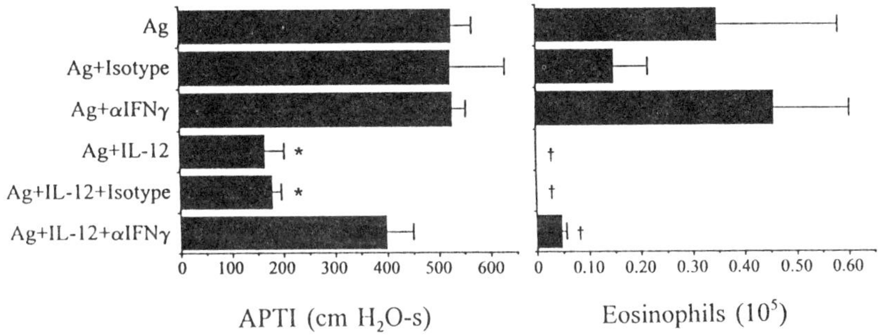

Figure 2. Dependence of inhibitory IL-12 effects on IFN-γ. Mice were sensitized i.p. with 10^8 SRBC in 0.5 ml PBS or PBS alone (Sham). Two weeks later, mice were intratracheally challenged with 6 x 10^8 SRBC in 0.05 ml PBS. SRBC-challenged mice were injected with vehicle (mouse serum) or 1 ug IL-12 i.p. for 5 consecutive days beginning 2 days before challenge, and measurements were determined 3 days after challenge. Some groups were injected i.p. with 2 mg isotype-matched control mAb or anti-IFN-γ mAb -2 days and +1 day relative to challenge. Airway pressure changes in response to i.v. administration of acetylcholine were measured and integrated over a 5 minute period and referred to as the airway pressure time index (APTI) in cm-H20-s. Results shown are mean +/- SE of three to six mice per group. * P <0.05 compared with antigen-challenged non-IL-12-treated groups and IL-12 + anti-IFN-γ treated group; †0.05 compared with all except antigen-only group; § 0.05 compared with all except antigen + isotype and antigen + anti-IFN-γ groups.

We provided further evidence for the importance of IL-12 in prevention of antigen-induced allergic airway responses by the observation that blockade of endogenous production of IL-12 in naturally resistant murine strains (C3H/HeJ) renders them susceptible to the development of allergen-induced airway hyperresponsiveness and eosinophilic inflammation [Keane-Myers *et al*, 1998] (Fig. 3). These results suggest that dysregulation of endogenous

IL-12 levels may be an important mechanism governing the pathogenesis of allergic airway disorders.

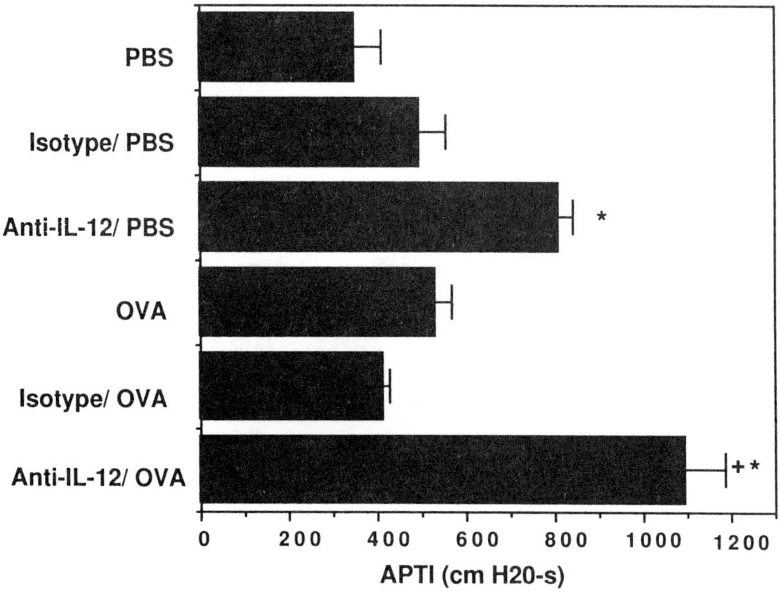

Figure 3. Interleukin-12 blockade renders naturally resistant C3H/HeJ mice susceptible to antigen-induced changes in airway reactivity. Mice were i.p. sensitized with 10ug ovalbumin in 50ul or an equivalent volume of PBS two weeks prior to challenge with either a 1.5% solution of ovalbumin or an equivalent volume of PBS. Four days after the challenge, mice are anesthesized and airway pressure changes in response to i.v. administration of acetylcholine (25ug/kg) were measured. Two days prior to antigen challenge mice were given anti-IL-12 mAb (1 mg) or an isotype control IgG2a (1 mg) in a volume of 200ul by i.p. injection. Airway pressure changes were integrated over a 5 minute period and referred to as the airway pressure time index (APTI) in cm-H20--sec Groups are designated as (antibody administration/sensitization and challenge). * denotes values significantly different from PBS groups ($P<0.05$); + ($P<0.05$) denotes values significantly different from those of OVA group. N= 8 mice/group.

Interestingly, recent examination of IL-12 levels in bronchial biopsies of asthmatic patients have shown that the number of cells positive for IL-12 p40 mRNA is significantly lower in asthmatic patients than in normal controls [Naseer *et al* 1997]. Furthermore, successful steroid treatment was accompanied by a significant increase in the number of cells expressing

IL-12 mRNA, whereas steroid therapy in steroid resistant patients did not result in an increase in IL-12 expressing cells. Additionally, another group has shown that Staphylococcus aureus (SAC)-induced production of IL-12 p70 in whole blood cultures from asthmatic patients was significantly reduced in comparison with nonatopic control subjects [van der Pouw Kraan *et al* 1997]. Taken together these studies suggest that deficient IL-12 production may predispose individuals to allergic asthma.

The mechanisms which give rise to alterations in IL-12 production are unclear at the present time, however several potential mechanisms exist. First alterations in IL-12 receptor expression or intracellular signalling in T cells may be responsible for expansion of type-2 cytokines. IL-12-induced responses have been shown to be mediated via induction of tyrosine phosphorylation of the transcription factor Stat4 [Jacobson *et al* 1995; Thierfelder *et al* 1996; Kaplan *et al* 1996]. Stat4 dimerizes and translocates to the nucleus where it binds to specific DNA sequences leading to IFN-γ production. Recently, two IL-12R subunits, termed $\beta 1$ and $\beta 2$ have been identified [Presky *et al* 1996]. IL-12R $\beta 1$ or $\beta 2$, independently expressed in COS cells, bind human IL-12 with low affinity, whereas coexpression of IL-12R $\beta 1$ and $\beta 2$ gives rise to high affinity binding sites. It is currently thought that IL-12 R$\beta 1$ is the signalling component of the IL-12R as it contains tyrosine residues in its cytoplasmic domain. Recent studies have shown that IL-12 responsiveness is loss in Th2 clones [Szabo *et al* 1995]. Consistent with this hypothesis, IL-12-induced activation of Stat4 only occurred in human Th1, but not Th2 cells [Rogge *et al* 1997]. Furthermore, IL-12 dependent signalling in human Th1 cells was shown to correlate with the selective expression of the transcripts encoding the signalling component of the IL-12 receptor (IL-12R) $\beta 2$ and with the presence of high affinity IL-12 binding sites selectively on Th1 cells. IL-12, and type 1 interferons induce expression of the IL-12R $\beta 2$ mRNAs, whereas IL-12R $\beta 2$ transcripts could not be detected in cells stimulated in the presence of IL-4. Thus, it is proposed that not only is IL-12R $\beta 2$ a marker of Th1 cells, but that lack of expression of this receptor subunit may lead to a Th2 polarized immune response. To our knowledge IL-12 receptor expression has not been examined in human asthmatics, but taken together with the finding that IL-12 production is deficient in the lungs of human asthmatics, we may speculate that selective expression and regulation of the IL-12 receptor $\beta 2$ subunit may be an important mechanism by which aberrant Th2 cell differentiation occurs in allergic diseases.

Alternatively, the production of IL-12 is either positively (IFN-γ) [Trinchieri 1994; Ma *et al* 1996] or negatively regulated by a number of cytokines and mediators (IL-4, IL-10, and PGE_2) [D'Andrea *et al* 1995; Meyaard *et al* 1996; van der Pouw Kraan *et al* 1995]. Dysregulation of any one of these mediators may lead to altered IL-12 production. Clearly, IL-4, IL-10 and PGE_2 levels are elevated in allergic diseases, whereas IFN-γ

levels are reduced. Thus a better understanding of the mechanisms underlying dysregulation of IL-12 production and responsiveness awaits studies designed to examine IL-12 receptor expression and modulation of IL-12 production in asthma.

VI. CONCLUSION

In summary, susceptibility to the development of allergic airway responses in response to inhaled indoor allergens likely originates from an abnormal specific immune response to specific indoor allergens such as HDM and animal dander. The T cells of asthmatic patients produce a type-2 cytokine profile namely IL-4 and IL-5 in response to allergen provocation, while normals presumably mount a type-1 immune response. This type of cytokine profile in the lung results in the characteristic eosinophilic inflammatory response and elevations in serum IgE levels. Although the mechanisms underlying this aberrant cytokine production are unclear, current studies suggest that altered regulation of either IL-4 or IL-12 production may lead to the expansion of deleterious Th2 cells in response to otherwise harmless inhaled antigens. Combined genetic, immunological and physiological approaches to this problem should result in a better understanding of these pathological processes and lead to development of therapeutics which may enable modulation of the immune response in this chronic disorder.

Acknowledgements

This work was supported by a contract from the Center for Indoor Air Research and by grant RO1HL58527 from the National Heart, Lung, and Blood Institute to M. Wills-Karp.

REFERENCES

Aikawa, T., S. Shimura, S. Hidetada, M. Ebina, and T. Takishima. 1992. Marked goblet cell hyperplasia with mucus accumulation in the airways of patients who died of severe acute asthma attack. *Chest*. 101:916-921.

Azzawi, M., B. Bradley, P. K. Jeffery, A. J. Frew, B. Assoufi, J. V. Collins, S. Durham, and A. B. Kay. 1990. Identification of activated T lymphocytes and eosinophils in bronchial biopsies in stable atopic asthmatics. *Am. Rev. Respir. Dis*. 142:1407-1413.

Beasley, R., W.R. Roche, J.A. Roberts, and S.T. Holgate. 1989. Cellular events in the bronchi in mild asthma and after bronchial provocation.

Am. Rev. Respir. Dis. 139:806-17.

Bousquet, J., P. Chanez, J.Y. Lacoste, G. Barneon, N. Ghavanian, I. Enander, P. Verge, S. Ahlstedt, J. Simony-La Fontaine, P. Godard, and F.B. Michel. 1990. Eosinophilic inflammation in asthma. *N. Eng. J. Med.* 323:1033-1039.

Bradding, P., I.H. Feather, P.H. Howarth, R. Mueller, J.A. Roberts, K. Britten, J.P.A. Bews, T.C. Hunt, Y. Okayama, C.H. Huesser. 1992. Interleukin 4 is localized to and released by human mast cells. *J. Exp. Med.* 176:1381.

Brown, D.R., D.J. Fowell, D. B. Corry, T.A. Wynn, N.H. Moskowitz, A.W. Cheever, R.M. Locksley, and S.L. Reiner. 1996. ß-2 microglobulin-dependent NK1.1+ T cells are not essential for T helper 2 immune responses. *J. Exp. Med.* 184:1295-1304.

Brusselle, G., J. Kips, G. Joos, H. Bluethmann, and R. Pauwels. 1995. Allergen-induced airway inflammation and bronchial responsiveness in wild-type and interleukin-4-deficient mice. *Am. J. Respir. Cell Mol.* 12:254-259.

Campbell H.D., W.Q. Tucker, Y. Hort, M.E. Martinson, G. Mayo, E.J. Clutterbuck, C.J. Sanderson, and I.G. Young. 1987. Molecular cloning, nucleotide sequence, and expression of the gene encoding human eosinophil differentiation factor (interleukin 5). *Proc. Natl. Acad. Sci. USA.* 84:6629-6633.

Clerici M., F.T. Hakim, D.J. Venzon, S. Blatt, C.W. Hendrix, T.A. Wynn, and G.M. Shearer. 1993. Changes in interleukin-2 and interleukin-4 production in asymptomatic human immunodeficiency virus-seropositive individuals. *J. Clin. Invest.* 91:759-765.

Cockcroft, D.W., D.N. Killian, J.J. Mellon, F.E. Hargreave 1977. Bronchial reactivity to inhaled histamine: A method and clinical survey. *Clin. Allergy.* 7:235-243.

Cockcroft, D.W. 1989. Airway hyperresponsiveness and late asthmatic responses. *Chest.* 94:178-80.

Cockcroft, D.W., K.Y. Murdock, and B.A. Berscheid. 1984. Relationship between atopy and bronchial responsiveness to histamine in a random population. *Ann. Allergy.* 53:26-29.

Corrigan C. J., and A. B. Kay. 1990. CD4+ T-lymphocyte activation in acute severe asthma: relationship to disease severity and atopic status. *Am. Rev. Respir. Dis.* 141:970-977.

Corry, D.B., H. G. Folkesson, M. L. Warnock, D. J. Erle, M. D. Matthay, J. P. Wiener-Kronish, and R. M. Locksley. 1996. Interleukin 4, but not interleukin 5 or eosinophils, is required in a murine model of acute airway hyperreactivity. *J. Exp. Med.* 183: 109-117.

Coyle, A. J., G. Le Gros, C. Bertrand, S. Tsuyuki, C. H. Heusser, M. Kopf and G. P. Anderson. 1995. Interleukin-4 is required for the induction of lung Th2 mucosal immunity. *Am. J. Respir. Cell. Mol. Biol.* 13:54-59.

Davydov, I.V., P. H. Krammer, and M. Li-Weber. 1995. Nuclear-factor-IL-6 activates the human IL-4 promoter in T cells. *J. Immunol.* 155:5273-5279.

De Monchy, J.G.R., H.F. Kauffman, P. Venge et al. 1985. Bronchoalveolar eosinophilia during allergen-induced late asthmatic reactions. *Am. Rev. Respir. Dis.* 131:373-6.

Del Prete G.F., M. de Carli, C. Mastromauro, R. Biagiotti, D. Macchia, P. Falagiani, M. Ricci, and S. Romagnani. 1991. Purified protein derivative of *Mycobacterium tuberculosis* and ex-cretory-secretory antigen(s) of *Toxocara canis* expand in vitro human T cells with stable and opposite (type 1 T helper or type 2 T helper) profile of cytokine production. *J. Clin. Invest.* 88:346-350.

DiCismo, B.F., G.P. Geba, D. Picarella, J.A. Elias, J.A. Rankin, B.R. Stripp, J.A. Whitsett, and R.A. Flavell. 1995. Airway epithelial cell expression of interleukin-6 in transgenic mice:uncoupling of airway inflammation and bronchial hyperreactivity. *J. Clin. Invest.* 94: 2028-2035.

D'Andrea, A., X. Ma, M. Aste-amezaga, C. Paganin, and G. Trinchieri. 1995. Stimulatory and inhibitory effects of interleukin (IL)-4 and IL-13 on the production of cytokines by human peripheral blood mononuclear cells: Priming for IL-12 and tumor necrosis factor-α production. *J. Exp. Med.* 181:537-46.

Finkelman, F.D., I.M. Katona, J.F. Urban, J. Holmes, J. Ohara, A.S. Tung, J.V. Sample, and W.E. Paul. 1988. IL-4 is required to generate and sustain in vivo IgE responses. *J. Immunol.* 141:2335-2341.

Foster, P. S., S. P. Hogan, A. J. Ramsay, K. I. Matthaei, and I. G. Young. 1996. Interleukin 5 deficiency abolishes eosinophilia, airways hyperreactivity, and lung damage in a mouse asthma model. *J. Exp. Med.* 183:195-201.

Frigas, E., D. A. Loegering, and G. J. Gleich. 1980. Cytotoxic effects of the guinea pig derived major basic protein on tracheal epithelium. *Lab. Invest.* 42:35-43.

Gavett, S.H., X. Chen, F. Finkelman, and M. Wills-Karp. 1994. Depletion of murine CD4+ T-lymphocytes prevents antigen-induced airway hyperreactivity and pulmonary eosinophilia. *Am. J. Cell. Mol. Biol.* 10:587-593.

Gavett, S. H., D. J. O'Hearn, C. L. Karp, E. A. Patel, B. H. Schofield, F. D. Finkelman, and M. Wills-Karp. 1997. Interleukin-4 receptor blockade prevents airway responses induced by antigen challenge in mice. *Am. J. Physiol. 272 (Lung Cell. Mol. Physiol.* 16):L253-L261.

Gavett, S.H., D. O'Hearn, X. Li, S. Huang, F.D. Finkelman, and M. Wills-Karp. 1995. Interleukin 12 inhibits antigen-induced airway hyperresponsiveness, inflammation and Th2 cytokine expression in mice. *J. Exp. Med.* 182: 1-10.

Gerblich, A. A., H. Salik, and M. R. Schuyler. 1991. Dynamic T-cell changes in peripheral blood and bronchoalveolar lavage after antigen bronchoprovocation in asthmatics. *Am. Rev. Respir. Dis.* 143:533-537.

Gereard, J., D. Rao, N. Morton. 1978. A genetic study of immunoglobulin E. *Am. J. Hum. Genet.* 30:46-58.

Gleich, G.J. 1990. The eosinophil and bronchial asthma: current understanding. *J. Allergy Clin. Immunol.* 85:422-436.

Gundel, R.H., L.G. Letts, and G.J. Gleich. 1991. Human eosinophil major basic protein induces airway constriction and airway hyperresponsiveness in primates. *J. Clin. Invest.* 87:1470-1473.

Hargreave, F.E., J. Dolovich, P.M. O'Byrne, E.H. Ramsdale, and E.E. Daniel. 1986. The origin of airway hyperresponsiveness, *J. Allergy Clin. Immunol.* 78:825-32.

Hershey, G. K. K., M. F. Friedrich, L. A. Esswein, M. L. Thomas, and T. A. Chatila. 1997. The association of atopy with a gain-of-function mutation in the a subunit of the interleukin-4 receptor. *N. Eng. J. Med.* 337:1720-5.

Hsieh, C.S., S.E. Macatonia, C.S. Tripp, S.F. Wolf, A. O'Garra, and K.M. Murphy. 1993. Development of Th1 CD4+ T cells through IL-12 produced by Listeria-induced macrophages. *Science.* 260:547-549.

Hsieh, C.S., A.B. Heimberger, J.S. Gold, A. O'Garra, and K.M. Murphy. 1992. Differential regulation of T helper phenotype development by interleukins 4 and 10 in an ab T-cell receptor transgenic system. *Proc. Natl. Acad. Sci. USA.* 89:6065-6069.

Jacobson, N. G., S. J. Szabo, R. N. Weber-Nordt, Z. Zhang, R. D. Schreiber, J. E. Darnell, and K. M. Murphy. 1995. Interleukin 12 signaling in T helper type 1 (Th1) cells involves tyrosine phosphorylation of signal transducer and activator of transcription (Stat) 3 and 4. *J. Exp. Med.* 181:1755-1762.

Kaplan, M. H., Y. -L. Sun, T. Hoey, and M. J. Grusby. 1996. Impaired IL-12 responses and enhanced development of Th2 cells in Stat4-deficient mice. *Nature (Lond.).* 382:174-177.

Kaplan, M. H., U. Schindler, S. T. Smiley, and M. J. Grusby. 1996. Stat6 is required for mediating responses to Il-4 and for the development of Th2 cells. *Immunity.* 4:313-319.

Kapsenberg M.L., E.A. Wierenga, J.D. Bos, and H.M. Jansen. 1991. Functional subsets of allergen-reactive human CD4+ T cells. *Immunol. Today* 12:392-395.

Katona, I. M., J.F. Urban, Jr., F.D. Finkelman, W.C. Gause, and K. B. Madden. 1993. Cytokine regulation of intestinal mastocytosis in Nippostrongylus brasiliensis infection. *Adv. In Exp. Med. Biol.* 371B:971-3.

Keane-Myers, M., M. Wysocka, G. Trinchieri, and M. Wills-Karp. 1998. Resistance to antigen-induced airway hyperresponsiveness requires endogenous production of interleukin-12. *J. Immunol.* 161:919-926.

Keegan, A.D., K. Nelms, L. Wang, J. H. Pierce, and W. E. Paul. 1994. Interleukin-4 receptor: signalling mechanisms. *Immunol. Today.* 15: 423-431.

Kips, J. C., G. J. Bursselle, G. F. Joos, R. A. Peleman, J. H. Tavernier, R. R. Devos, and R. A. Pauwels. 1996. Interleukin-12 inhibits antigen-induced airway hyperresponsiveness in mice. *Am. J. Respir. Crit. Care Med.* 153:535-539.

Korsgren, M., J. S. Erjefalt, O. Korsgren, F. Sundler, C. G. A. Persson. 1996. Allergic eosinophil-rich inflammation develops in lungs and airways of B cell-deficient mice. *J. Exp. Med.* 185:885-892.

Kraneveld, A. D., F. P. Nijkamp, and A. J. M. Van Oosterhout. 1997. Role of neurokinin-2 receptor in interleukin-5-induced airway hyperresponsiveness but not eosinophilia in guinea pigs. *Am. J. Respir. Crit. Care Med.* 156:367-374.

Kung, T. T., D. M. Stelts, J. A. Zurcher, G. K. Adams, III, R. W. Egan, W. Kreutner, A. S. Watnick, H. Jones, and R. W. Chapman. 1995. Involvement of IL-5 in a murine model of allergic pulmonary inflammation: Prophylactic and therapeutic effect of an anti-IL-5 antibody. *Am. J. Respir. Cell Mol. Biol.* 13:360-365.

Kuperman, D., B. Schofield, M. Wills-Karp, and M. J. Grusby. 1998. Signal transducer and activator of transcription factor 6 (Stat6)-deficient mice are protected from antigen-induced airway hyperresponsiveness and mucus production. *J. Exp. Med.* 187:939-948.

Launois, P., K.G. Swihart, G. Milon, and J.A. Louis. 1997. Early production of IL-4 in susceptible mice infected with Leishmania major rapidly induced IL-12 unresponsiveness. *J. Immunol.* 158:3317-3324.

Lefort, J., C. Bachelet, D. Leduc, B. B. Vargaftig. 1996. Effects of antigen provocation of IL-5 transgenic mice on eosinophil mobilization and bronchial hyperresponsiveness. *J. Allergy Clin. Immunol.* 97:788-799.

Lin, J. -X, T. -S. Migone, M. Friedman, J. A. Weatherbee, L. Zhou, A. Yamauchi, E. T. Bloom, J. Mietz, S. John, and W. J. Leonard. 1995. The role of shared receptor motifs and common Stat proteins in the generation of cytokine pleiotrophy and redundancy by IL-2, IL-4, IL-7, IL-13, and IL-15. *Immunity.* 2:331-339.

Locksley, R.M. 1994. Th2 cells: Help for helminths. *J. Exp. Med.* 179:1405-1407.

Lopez, A. F., C. J. Sanderson, J. R. Gamble, H. R. Campbell, I. G. Young, and M. A. Vadas. 1988. Recombinant human interleukin-5 is a selective activator of human eosinophil function. *J. Exp. Med.* 167:219-224.

Lundgren, R., M. Soderberg, P. Horstedt, and R. Stenling. 1988.

Morphological studies of bronchial mucosal biopsies from asthmatics before and after ten years of treatment with inhaled steroids. *Eur. Respir. J.* 1:883-889.

Luoma, R., A. Koivikko, and M. Viander. 1983. Development of asthma, allergic rhinitis and atopic dermatitis by the age of five years. *Allergy.* 38:339-46.

Ma, X., J.M. Chow, G. Gri, G. Carra, F. Gerosa, S.F. Wolf, R. Dzialo, and G. Trinchieri. 1996. The interleukin-12 gene promoter is primed by interferon-gamma in monocytic cells. *J. Exp. Med.* 183:147-57.

Madden, K.B., J.F. Urban, Jr., H.J. Ziltener, J.W. Schrader, F.D. Finkelman,and I.M. Katona. 1991. Antibodies to IL-3 and IL-4 suppress helminth-induced intestinal mastocytosis. *J. Immunol.* 147:1387-1391.

Marsh, D.G., Neely, J.D., D.R. Breazeale, B. Ghosh, L.R. Freidhoff, E. Ehrlich-Kautzky, C. Schou, G. Krishnaswany, and T.H. Beaty. 1994. Linkage analysis of IL4 and other chromsome 5q31.1 markers and total serum immunoglobulin E concentrations. *Science.* 264:1152-1156.

Martinez, F. D., D. A. Stern, A. L. Wright, C. J. Holberg, L. M. Taussig, and M. Halonen. 1995. Association of interleukin-2 and interferon-gamma production by blood mononuclear cells in infancy with parental allergy skin tests and with subsequent development of atopy. *J. Allergy Clin. Immunol.* 96:652-60.

Mattoli, S., V.L. Mattoso, M. Saloperto, L. Allegra, A. Fasoli. 1991. Cellular & biochemical characteristics of bronchoalveolar lavage fluid in symptomatic & nonallergic asthma. *J. Allergy Clin. Immunol.* 87: 794-802.

McFadden, E. R., and I. A. Gilbert. 1992. Asthma. *N. Engl. J. Med.* 327:1928-1937.

Melhop, P. D., M. Van de Rijn, A. B. Goldberg, J. P. Brewer, V. P. Kurup, T. R. Martin, and H. C. Oettgen. 1997. Allergen-induced bronchial hyperreactivity & eosinophilic inflammation occur in the absence of IgE in mouse model of asthma. *Proc. Natl. Acad. Sci.* 94:1344-49.

Meyaard, L., E. Hovenkamp, S.A. Otto, and F. Miedema. 1996. IL-12-induced IL-10 production by human T cells as a negative feedback for IL-12-induced immune responses. *J. Immunol.* 156:2776-2782.

Mochizuki, M., J. Bartels, A. I. Mallet, E. Christophers, and J. M. Schroder. 1998. IL-4 induces eotaxin: a possible mechanism of selective eosinophil recruitment in helminth infection and atopy. *J. Immunol.* 160:60-68.

Mosmann, T. R., H. Cherwinski, M. W. Bond, M. A. Gieldin, and R.L. Coffman. 1986. Two types of murine helper T cell clone. I. Definition according to profiles of lymphokine activities and secreted proteins. *J. Immunol.* 136:2348-2357.

Nakajima, H., I. Iwamoto, S. Tomoe, R. Matsumaria, H. Tomioka, K.

Takatsu, and S. Yoshida. 1992. CD4+ T lymphocytes and interleukin-5 mediate antigen-induced eosinophil infiltration into mouse trachea. *Am. Rev. Respir. Dis.* 146:374-377.

Nakajima, H., H. Sano, T. Nishimura, S. Yoshida, and I. Iwamoto. 1994. Role of vascular cell adhesion molecule 1/very late activation antigen 4 and intercellular adhesion molecule 1/lymphocyte function-associated antigen 1 interactions in antigen-induced eosinophil and T cell recruitment into the tissue. *J. Exp. Med.* 170:1145-1154.

Naseer, T., E.M. Minshall, D. Y. M. Leung, S. Laberge, P. Ernst, R.J. Martin, and Q. Hamid. 1997. Expression of IL-12 and IL-13 mRNA in asthma and their modulation in response to steroid therapy. *Am. J. Respir. Crit. Care Med.* 155:845-851.

Olivenstein, R., P. M. Renzi, L. J. Xu, J. P. Yang, and J. G. Martin. 1994. Effects of W3/25 monoclonal antibody on pulmonary inflammation and the late airway response in Brown-Norway rats. *Am. Rev. Respir. Dis.* 149:528a.

Pauwels, R. 1989. The relationship between airway inflammation and bronchial hyperresponsiveness. *Clinical and Exp. Allergy.* 19:395-398.

Peltz G. 1991. A role for CD4+ T-cell subsets producing a selective pattern of lymphokines in the pathogenesis of human chronic inflammatory and allergic diseases. *Immunol. Rev.* 123:23-35.

Platts-Mills, T.A.E., E.R. Tovey, E.B. Mitchell, H. Moszorz, P. Nock, S.R. Wilkins. 1982. Reduction of bronchial hyperreactivity during prolonged allergen avoidance. *Lancet.* ii:675-678.

Postma, D. S., E. R. Bleecker, P. J. Amelung, K. J. Holroyd, J. Xu, C. I. Panhuysen, D. A. Meyers, and R. C. Levitt. 1995. Genetic susceptibility to asthma: bronchial hyperresponsiveness coinherited with a major gene for atopy. *N. Eng. J. Med.* 333:894-900.

Presky, D. H., H. Yang, L. J. Minetti, A. O. Chua, N. Nabavi, C. -Y Wu, M. K. Gately, and U. Gubler. 1996. A functional interleukin-12 receptor complex is composed of two B type cytokine receptor subunits. *Proc. Natl. Acad. Sci. USA.* 93:14002-14007.

Pretolani, M., C. Ruffie, J.-R. Lapa e Silva, D. Joseph, R. R. Lobb, and B. B. Vargaftig. 1994. Antibody to very late activation antigen 4 prevents antigen-induced bronchial hyperreactivity and cellular infiltration in the guinea pig airways. *J. Exp. Med.* 180:795-805.

Rankin, J. A., D. E. Picarella, G. P. Geba, U. Temann, B. Prasad, B. DiCosmo, A. Tarallo, B. Stripp, J. Whitsett, and R. A. Flavell. 1996. Phenotypic and physiologic characterization of transgenic mice expressing interleukin 4 in the lung: lymphocytic and eosinophilic inflammation without airway hyperreactivity. *Proc. Natl. Acad. Sci. USA.* 93:7821-7825.

Rincon, M., J. Anguita, T. Nakamura, E. Fikrig, and R.A. Flavell. 1997. Interleukin (IL)-6 directs the differentiation of IL-4-producing CD4+ T cells. *J. Exp. Med.* 185:461-469.

Robinson, D. S., Q. Hamid, S. Ying, A. Tsicopoulos, J. Barkans, A. M. Bentley, C. Corrigan, S. R. Durham, and A. B. Kay. 1992. Predominant TH2-like bronchoalveolar T-lymphocyte population in atopic asthma. *N. Engl. J. Med.* 326:298-304.

Rogge, L., L. Barneris-Maino, M. Biffi, N. Passini, D. H. Presky, U. Gubler, and F. Sinigaglia. 1997. Selective expression of an interleukin-12 receptor component by human T helper 1 cells. *J. Exp. Med.* 185:825-831.

Romagnani, S. 1994. Lymphokine production by human T cells in disease states. *Annu. Rev. Immunol.* 12:227.

Savoie, C., M. Plant, M. Zwikker, C. J. Van Staden, L. Boulet, C. C. Chan, I. W. Rodger, and D. J. Pon. 1995. Effects of dexamethasone on antigen-induced high molecular weight glycoconjugate secretion in allergic guinea pigs. *Am. J. Respir. Cell Mol. Biol.* 13:133-143.

Schleimer, R. P., S. Sterbinsky, and S. Kaiswer. 1992. IL-4 induced adherence of human eosinophils and basophils but not neutrophils to endothelium. Association with expression of VCAM-1. *J. Immunol.* 148:1086-1092.

Schmitz, J., A. Thiel, R. Kuhn, K. Rajewsky, W. Muller, M. Assenmacher, and A. Radbruch. 1994. Induction of interleukin 4 (IL-4) expression in T helper (Th) cells is not dependent on IL-4 from non-Th cells. *J. Exp.Med.* 179:1349-1353.

Sears, M.R., B. Burrows, E.M. Flannery, G. Herbison, C.J. Hewitt, and M.D. Hodaway. 1991. Relation between airway responsiveness and serum IgE in children with asthma and in apparently normal children. *N. Eng. J. Med.* 325:1067-1071.

Sears, M.R., G.P. Herbison, M.D. Holdaway, C.J. Hewitt, E.M. Flannery, and P.A. Silva. 1989. The relative risks of sensitivity to grass pollen, house dust mite and cat dander in the development of childhood asthma. *Clin. Exp. Allergy.* 19:419-424.

Seder, R.A., W.E. Paul, M.M. Davis, and B. Fazekas de St Groth. 1992. The presence of interleukin 4 in vitro priming determines the lymphokine-producing potential of CD4+ T cells from T cell receptor transgenic mice. *J. Exp. Med.* 176:1091-1098.

Seder, R.A., W.E. Paul, A.M. Dvorak, S.J. Sharkis, A. Kagey-Sobotka, Y. Niv, F.D. Finkelman, S.A. Barnieri, S.J. Galli, and M. Plaut. 1991. Mouse splenic and bone marrow cell populations that express high-affinity FC receptors and produce interleukin 4 are highly enriched in basophils. *Proc. Natl. Acad. Sci.* USA. 88:2835.

Sher A., D. Fiorentino, P. Caspar, E. Pearce and T.R. Mosmann.1991. Production of IL-10 by CD4+ lymphocytes correlates with down-regulation of Th1 cytokine synthesis in helminth infections. *J. Immunol.* 147:2713-2716.

Shirakawa, T., T. Enomoto, S. Shimazu, and J.M. Hopkin. 1997. The inverse association between tuberculin responses and atopic disorder. *Science.* 275:77-79.

Stenius, B., L. Wide, W.M. Seymour, V. Holford-Strevens, and J. Pepys. 1971. Clinical significance of specific IgE to common allergens. *Clin. Allergy.* 1:37-55.

Street N.E., and T.R. Mosmann. 1991. Functional diversity of T-lymphocytes due to secretion of different cytokine patterns. *FASEB.* 5:171-177.

Sun, X. L., L. -M. Wang, Y. Zhang, L. Yenush, M. G. Myers, Jr., E. Glasheen, W. S. Lane, J. H. Pierce, and M. F. White. 1995. Role of IRS-2 in insulin and cytokine signalling. *Nature.* 377:173-177.

Swain, S. L., A. D. Weinberg, M. English, and G. Hutson. 1990. IL-4 directs the development of Th2-like helper effectors. *J. Immunol.* 145:3796-3806.

Szabo, S. J., N. G. Jacobson, A. S. Dighe, U. Gubler, and K. M. Murphy. 1995. Developmental commitment to the Th2 lineage by extinction of IL-12 signalling. *Immunity.* 2:665-675.

Temann, U. A., B. Prasad, M. W. Gallup, C. Basbaum, S. B. Ho, R. A. Flavell, and J. A. Rankin. 1997. A novel role for murine IL-4 in vivo: induction of MUC5AC gene expression and mucin hypersecretion. *Am. J. Respir. Cell Mol. Biol.* 16:471-478.

Thierfelder, W. E., J. M. Van Deursen, K. Yamamoto, R. A. Tripp, S. R. Sarawar, R. T. Carson, M. Y. Sangster, D. A. A. Vignali, P. C. Doherty, G. C. Grosveld, and J. N. Ihle. 1996. Requirement for Stat4 in interleukin-12-mediated responses of natural killer and T cells. *Nature (Lond).* 382:171-174.

Trinchieri G. 1994. Interleukin-12: A cytokine produced by antigen-presenting cells with immunoregulatory functions in the generation of T-helper cells type 1 and cytotoxic lymphocytes. *Blood.* 84:4008-27.

Van der Pouw Kraan TCTM, L.C.M. Boeije, R.J.T. Smeenk, J. Wijdenes, and L.A. Aarden. 1995. Prostaglandin-E2 is a potent inhibitor of human interleukin 12 production. *J. Exp. Med.* 181:775-779.

Van der Pouw Kraan, TCTM, L.C. M. Boeije, and E. R. De Groot, 1997. Reduced production of IL-12 and IL-12-dependent IFN-γ release in patients with allergic asthma. *J. Immunol.* 158:5560-5565.

Van Snick, J. 1990. Interleukin-6: an overview. *Annu. Rev. Immunol.* 8:253-278.

Van Oosterhout, A. J., A. R. Ladenius, H. F. Savelkoul, I. Van Ark, K. C. Delsman, and F. P. Nijkamp. 1993. Effect of anti-IL-5 and IL-5 on airway hyperreactivity and eosinophils in guinea pigs. *Am. Rev. Respir. Dis.* 147:548-552.

Walker, C., E. Bode, L. Boer, T. T. Hansel, K. Blaser, and J. C. Virchow, Jr. 1992. Allergic and nonallergic asthmatics have distinct patterns of T-cell activation and cytokine production in peripheral blood and bronchoalveolar lavage. *Am. Rev. Respir. Dis.* 146:109-115.

Wang, J. M., A. Rambaldi, A. Biondi, Z. G. Chen, C. J. Sanderson, and A. Mantovani. 1989. Recombinant human interleukin-5 is a selective eosinophil chemoattractant. *Eur. J. Immunol.* 19:701-705.

Wardlaw, A. J., S. Dunnette, G. J. Gleich, J. V. Collins, and A. B. Kay. 1988. Eosinophils and mast cells in bronchoalveolar lavage in subjects with mild asthma. Relationship to bronchial hyperreactivity. *Am. Rev. Respir. Dis.* 137:62-69.

Wardlaw A. J., S. Dunnette, G. J. Gleich, J. V. Collins, and A. B. Kay. 1988. Eosinophils and mast cells in bronchoalveolar lavage in subjects with asthma: relationship to bronchial hyperreactivity. *Am. Rev. Respir. Dis.* 137:62-69.

Weller, P. F., K. Lim, H. C. Wan, A. M. Dvorak, D. T. Wong, W. W. Cruikshank, H. Kornfeld, and D. W. Center. 1996. Role of the eosinophil in allergic reactions. *Eur. Respir. J.* 22:109s-115s.

Wilkinson, J.R., S.J. Lane, and T.H. Lee. 1991. The effects of corticosteroids on cytokine generation and expression of activation antigens by monocytes in bronchial asthma. *Int. Arch. Allergy Clin. Immunol.* 94:220-221.

Witthuhn, B. A., O. Silvennoinen, O. Miura, K. S. Lai, C. Cwik, E. Liu, and J. N. Ihle. 1994. Involvement of the Jak-3 Janus Kinase in signalling by interleukins 2 and 4 in lymphoid and myeloid cells. *Nature.* 370:153-157.

Yamamura M., K. Uyemura, R.J. Deans, K. Weiberg, T.H. Rea, B.R. Bloom, and R.L. Modlin. 1991. Defining protective responses to pathogens: cytokine profiles in leprosy lesions. *Science.* 254:277-279.

Yoshimoto, T., A. Bendelac, C. Watson, J. Hu-Lui, and W.E. Paul. 1995. Role of NK1.1+ T cells in a Th2 response and in immunoglobulin E production. *Science.* 270:1845-1847.

Yoshimoto, T., and W. E. Paul. 1994. CD4pos, NK1.1pos T cells promptly produce interleukin 4 in response to in vivo challenge with anti-CD3. *J. Exp. Med.* 179:1285-1295.

Chapter 6

MODIFIED RECOMBINANT ALLERGENS FOR IMMUNOTHERAPY

Alisa M. Smith, PhD

I. INTRODUCTION

Over the past 10 years, molecular cloning techniques have become established for the identification and characterization of allergen molecules. More than 20 indoor allergens, as well as many pollen and venom allergens, have been cloned and sequenced and allergens now constitute a large, well defined group of proteins for biomedical research (Thomas, 1996; Chapman et al., 1997; Scheiner and Kraft, 1995). Strategies for improving diagnosis of allergic disease and asthma therapy, based on recombinant allergens, have become feasible now that major allergens can be produced in large quantities, purified to homogeneity, and modified using genetic engineering.

In spite of the advances in characterization and standardization of allergen extracts, immunotherapy for allergic disease is practiced today much as it has been since 1911, when Noon and Freeman established the fundamental approach by administering pollen extracts for the treatment of seasonal hayfever (Noon, 1911; Freeman, 1911). Advances in our understanding of the immunologic events responsible for hypersensitivity suggest that the regulation of T cell responses to allergens is the key to successful immunotherapy. However, the precise mechanisms remain obscure (Wallner and Gefter, 1994). It is clear that the dose, the quality of the allergen extract and the length of the therapy are key to achieving effective treatment (Creticos et al. 1989). There are side effects associated with conventional immunotherapy which range from swelling at the site of injection to anaphylaxis. Currently, research is focussed on improving allergen specific immunotherapy by using recombinant DNA technology. The aims of this chapter are to review the current knowledge of the molecular biology of house dust mite allergens, one of the principle indoor allergens associated with asthma, and to outline different strategies for the use of recombinant allergens in immunotherapy.

II. MOLECULAR BIOLOGY OF HOUSE DUST MITE ALLERGENS

A. IDENTIFICATION AND CLONING

The Group 1 and Group 2 mite allergens had been identified by protein purification from extracts of mite culture media and shown to be important based on their reactivity with IgE Ab from the majority of dust mite allergic

1-56670-373-5/99/$0.00+$.50

patients. In 1988, Chua and Thomas cloned the 24kDa *Dermatophagoides pteronyssinus* allergen, Der p 1 (Chua et al., 1988; Thomas et al., 1988). Subsequently, they and others cloned Der p 2 and the Group 1 and Group 2 homologues from *D. farinae* (Chua et al., 1990; Dilworth et al., 1991; Yuuki et al., 1990; Trudinger et al., 1991). Today, cDNA cloning has defined the primary amino acid sequence of at least ten different groups of allergens from *Dermatophagoides spp.*, as well as several homologous allergens from *Euroglyphus, Lepidoglyphus* and *Blomia spp.* (Thomas, 1996; Chapman et al., 1997). The prevalence of IgE Ab binding to these proteins varies depending on the patient population, however, as many as six of these allergens react with IgE Ab from >50% of mite allergic patients and are considered to be important allergens (Table 1).

The biologic function of several allergens has been determined by protein sequence homology (Table 1). Der p 1, Der p 3 and Der p 9 are cysteine or serine proteases, and Der p 4 is an amylase. Based on the proteolytic activity of these allergens, Stewart and others have postulated that the enzyme activity of allergens influences the IgE antibody response, perhaps by facilitating allergen uptake by the respiratory mucosa or by cleavage of cell surface CD23, which causes an up-regulation of this IgE receptor (Robinson et al., 1997; Hewitt et al., 1995). Homology with proteins of known structure has been used to model the tertiary structure of allergens, for example Der p 1 has been modeled on the papain structure (Topham et al., 1994). Not only does this provide information about the biologic function of these allergens but structural models provide a picture of the antigenic surface recognized by Ab; information that is important for identifying antigenic determinants for IgE antibody and T cells.

Table 1. *Dermatophagoides* Allergens: Structure /Function

Allergen[1]	Prevalence of IgE Ab	MW (kDa)	Biologic Function
Group 1	90%	25	Cysteine protease
Group 2	90%	14	*Unknown* [2]
Group 3	~60%	30	Serine protease
Der p 4	40%	60	Amylase
Der p 5	40%	14	*Unknown*
Der p 6	~50%	25	Chymotrypsin
Group 7	50%	22-28	*Unknown*
Der p 8	40%	26	GSH-transferase
Der p 9	30%	24	Serine protease
Group 10	~50%	36	Tropomyosin

[1] Third International Workshop Report on Indoor Allergens in Asthma. J Allergy Clin Immunol '97;100:S1-S24.

[2] The Group 2 allergens show sequence homology to a mammalian reproductive protein. Thomas & Chua, Clin Exp Allergy '95;25:677.

B. EXPRESSION OF RECOMBINANT ALLERGENS

High level *in vitro* expression systems have been developed as a consequence of cloning studies. This has provided larger quantities and more pure preparations of allergens that can be readily standardized in mass units. Most commonly, recombinant allergens have been expressed in *E. coli* or in the yeast *Saccharomyces* and *Pichia*. For example, recombinant Der p 2 has been produced in bacteria with a final yield of 20mg/L and in *Saccharomyces* at 10 mg/L (Mueller et al., 1997, Hakkaart et al., 1997). Many of the recombinant allergens show comparable immunoreactivity to the natural allergens, measured using RAST, quantitative skin testing, and *in vitro* T cell proliferation assays. Recombinant Group 2, 5 and 7 allergens show an excellent correlation with natural allergen for IgE antibody binding using serum from patients with asthma, atopic dermatitis, and rhinitis and quantitative skin testing has confirmed this correlation (Smith and Chapman, 1996; Chapman et al., 1997, Shen et al., 1996). The Group 1 mite allergens have been an exception; recombinant Der p 1 produced in bacteria or yeast showed only 50% reactivity (IgE binding) compared to the natural allergen. However, Shoji et al. recently demonstrated that rDer f 1, with immunoreactivity and enzyme activity similar to the natural allergen, could be produced in insect cells using a baculovirus expression system (Shoji et al., 1996). Recombinant allergen expression systems have provided material for the development of sensitive immunoassays for detecting picogram quantities of allergen in house dust (Chapman et al., 1995). In addition, high level expression systems have provided material for structural studies. This is especially important for those allergens where primary amino acid analysis has failed to define the structure/function, such as the Group 2 and Group 5 mite allergens. The three dimensional structure of Der p 2 has been determined using nuclear magnetic resonance spectroscopy (NMR). The protein consists of two three-stranded (β-sheets in a novel arrangement (Mueller et al., 1997). This is the first structure of an important indoor allergen.

III. NEW STRATEGIES FOR IMMUNOTHERAPY

A. RATIONALE

Conventional immunotherapy using increasing doses of standardized extracts can be associated with adverse reactions including swelling and discomfort at the site of injection, local edema, itching of the eyes or nose, generalized hives, and more serious systemic reactions that may lead to anaphylaxis. It is possible that the frequency of adverse reactions associated with immunotherapy might decrease using recombinant allergens. This might occur because preparations would lack the non-allergenic components found in extracts and dosing could be better controlled. However, immunotherapy with recombinant allergens may produce side effects similar to those seen with the natural allergens, thus the rationale behind new strategies is to reduce IgE Ab reactivity with the aim of decreasing side effects while providing effective treatment. Table 2 summarizes several of these approaches.

Table 2. Asthma Immunotherapy: New Strategies

- Recombinant Allergens
- Allergen variants with reduced IgE reactivity
- Peptides from Allergens
- Allergens linked to adjuvants
- Plasmid DNA vaccines
- Anti-IgE therapy
- Anti-cytokine/receptor therapy
- Anti-sense DNA

B. MODIFIED ALLERGENS

Advances in our knowledge of T cell immunology include the characterization of the T cell receptor, MHC and peptide interactions, antigen processing, and cytokine production. This has focussed attention on the T cell compartment of immediate hypersensitivity as a prime target for interventive strategies (O'Hehir et al., 1991). In addition, synthetic peptide chemistry has made it possible to evaluate the T cell epitopes for any allergen of known primary amino acid sequence. The allergic response is dominated by T cells of the TH2 phenotype, secreting primarily IL-4 and IL-5, which are required for IgE Ab production and eosinophilia. Down regulating this TH2 response, or redirecting the response to the TH1 phenotype, is the focus of several studies.

Peptide Therapy. The notion of down regulating allergic responses by using peptides corresponding to T cell epitopes was supported by studies in murine systems demonstrating that T cells could be rendered non-responsive by immunization with high doses of a peptide representing an immuno-dominant T cell epitope (Lamb et al., 1983; Hoyne et al., 1993). This T cell anergy was characterized by functional changes including decreased IL-4 production as well as changes in cell surface molecules such as down-regulation of the T cell receptor and up-regulation of the IL-2 receptor, and increased turnover of CD28 transcripts (Hoyne et al., 1993; Lake et al., 1993). In most cases, peptides react poorly with IgE Ab, and thus peptide immunotherapy would be associated with fewer side effects. Peptides can be produced in large quantities, and this strategy does not depend on *in vitro* protein expression and purification systems.

Detailed studies on the use of T cell peptides have been carried out using the cat allergen Fel d 1. Stimulatory T cell epitopes are found throughout both Chain 1 and Chain 2 (van Neerven et al., 1994; Counsell et al., 1996). However, two large peptides from Chain 1 were identified that stimulated T cells from the majority of cat allergic subjects and an immunotherapy trial has recently been published (Norman et al., 1996). Those patients receiving the highest dose of peptide showed significantly fewer symptoms upon exposure than the placebo group ($p<0.05$).

It is important to note that although the peptides did not induce skin test reactivity prior to beginning the study, there were side effects in 16 of 24 patients receiving the highest dose. The side effects included chest tightness, runny nose, and itching without hives and these symptoms decreased with subsequent doses. It is also important to point out that the challenge used in this study was more rigorous than natural exposure conditions, making it difficult to equate the symptom improvement seen in this study with improvement seen during conventional immunotherapy. This strategy appears promising, in part because the injections were given over 4 weeks as opposed to standard immunotherapy which requires a much longer period to reach a maintenance dose. Recently, the results of a similar trial using three peptides from the ragweed allergen Amb a 1 have been reported (Creticos et al., 1997). Thus, peptide based immunotherapy may be effective for the treatment of cat and ragweed allergy for which one major allergen is responsible for inducing symptoms. These results raise the question "Can such a strategy be applied to treatment of dust mite allergy?" where three or more allergens stimulate IgE Ab and T cell responses.

The T cell response to Der p 1 and Der p 2 has been studied using population responses of peripheral blood lymphocytes as well as T cell clones from atopic and non-atopic subjects (Yssel et al., 1992; Higgins et al., 1994; O'Brien et al., 1994; van Neerven et al., 1993; O'Hehir et al., 1993; Verhoef et al., 1993). These studies have demonstrated that T cell epitopes can be found throughout the entire coding sequence of Der p 1 and Der p 2. However, regions of Der p 1, for example, recognized by the majority of allergic individuals were localized to three overlapping peptides spanning amino acids 107-131 (Higgins et al., 1994). Similarly, commonly recognized Der p 2 peptides were mapped to amino acids 11-35 and 61-104 (O'Brien et al., 1994; van Neerven et al., 1993). Studies by Hoyne and colleagues showed that Der p 1 sensitized mice treated intranasally with the peptide 111-139 were unresponsive to parenteral challenge with Der p 1 (Hoyne et al., 1993; Lake et al., 1993). An important conclusion from these studies was that the T cell response to the entire Der p 1 could be down regulated using a single peptide. It may thus be possible to prepare a mixture of peptides representing T cell epitopes of three or four mite allergens, for evaluation in clinical trials. A test of this hypothesis awaits further *in vitro* T cell studies to identify immunodominant T cell epitopes from additional mite allergens other than Group 1 and Group 2.

Allergen Variants. One drawback to peptide-based immunotherapy is the requirement for identifying immunogenic epitopes of a particular allergen that are recognized by T cells from the diverse HLA haplotypes of the allergic population. To date, no clear correlation exists between response to a major allergen and a specific HLA type (van Neerven et al., 1996). An alternative strategy involves using the entire allergen molecule containing a more complete repertoire of T cell epitopes, but in a form with reduced IgE Ab reactivity. Michael and colleagues produced peptic digests of short ragweed and compared the fragmented allergen preparation to a conventional extract in an immunotherapy trial. The digested allergen was well tolerated and the group of subjects receiving this form of the allergen showed greater improvement with respect to symptom scores and IgG levels, compared to the group receiving the conventional treatment (Litwin et al., 1992). Other strategies for modifying allergens include polymerization with glutaraldehyde or formaldehyde, or coating the allergen with a polymer such as polyethylene glycol (PEG) (Ishizaka et al., 1974; Guerra et al., 1993). These treatments are effective in reducing IgE reactivity of the allergens, and murine studies have shown that glutaraldehyde aggregated allergen can preferentially induce IFN-γ (Yang et al., 1993). However, the difficulty in controlling the extent of modification from batch to batch made these reagents not readily acceptable to regulatory agencies. Most recently, this idea has been approached using genetically engineered allergen variants. Site-specific mutagenesis *in vitro* of allergen cDNA can be used to change the amino acid sequence to produce precisely defined variants. For example, one or two amino acid substitutions in Der p 2 were produced using oligonucleotide-directed mutagenesis; these substitutions were targeted to alter cysteine residues, thus selectively disrupting one or two disulfide bonds. These variants showed reduced IgE Ab reactivity on skin testing by up to 100-fold (Smith and Chapman, 1996). Table 3 shows the results of intradermal skin tests comparing rDer p 2 with two variants. *In vitro* T cell proliferative responses to these variants were similar to proliferation using unmodified Der p 2, suggesting that these variants retain T cell epitopes (Smith et al., 1998). Such modified allergens could be given in doses similar to those used for conventional immunotherapy and patients would be expected to reach a maintenance dose more rapidly and with fewer side effects due to the decreased IgE Ab reactivity of the variants.

In a parallel study, Der p 2 variants generated at predicted surface residues also showed significant loss in IgE Ab reactivity (Smith and Chapman, 1997). Thus structural information is important for targeting amino acids for mutagenesis and extending this strategy to other allergens such as the Group 1 proteins. In some cases, naturally occurring variants have been identified that show decreased IgE Ab binding (e.g. Bet v 1) and this suggests an additional source of allergen variants suitable for immunotherapy (Ferreira et al., 1996).

Table 3.

Intradermal Skin Test Responses to rDer p 2 and Cysteine Variants[1]

Patient	Antigen	Wheal Size (mm) at Allergen Concentration (μg/ml)				
		10^{-4}	10^{-3}	10^{-2}	10^{-1}	10^{0}
	rDer p 2	-	8 x 7	10 x 7		
M.P.	Cys73	-	*	8 x 6	10 x 8	
	Cys21+73	-	-	-	*	5 x 4
	rDer p 2	*	9 x 9			
A.A.	Cys73	-	7 x 6	8 x 8		
	Cys21+73	-	-	*	*	8 x 5
	rDer p 2	-	7 x 7	10 x 7		
G.M.	Cys73	-	*	5 x 7	10 x 9	
	Cys21+73	-	-	-	*	7 x 5

[1] Reaction was measured 15 min after intradermal injection of 0.02 ml of allergen.
- denotes reaction indistinguishable from diluent control;
* denotes flare without significant wheal.

C. OTHER STRATEGIES

Adjuvants and Prophylactic Vaccines. Adjuvants are included in conventional vaccine strategies and it is thought that they promote an immune response by slowing the release of antigen, targeting antigen to macrophages, and perhaps specifically activating macrophages to produce IL-12. Murine models of Leishmaniasis have shown that IL-12 can be used in place of bacterial adjuvants in promoting a cell-mediated immune response to Leishmania antigens (Julia et al., 1996). In a murine model of airway hyper-responsiveness, administration of IL-12 at the time of antigen challenge abolished pulmonary eosinophilia and hyperreactivity in sensitized animals (Gavett et al., 1995).

Holt and colleagues have proposed vaccination of children of atopic parents using a prophylactic vaccine composed of allergen with a TH1 adjuvant (Holt, 1994). Their hypothesis is that this type of immunization would promote a TH1/IgG antibody response that would be protective and would prevent subsequent TH2/IgE antibody responses. Although there have been no clinical studies of prophylactic vaccines for asthma, there are parallels with other pediatric vaccines such as H. influenza conjugate vaccines (Frasch, 1995).

Naked DNA Vaccines. DNA-based vaccine strategies for the treatment of airway hyper-responsiveness (AHR) have been investigated in animal model systems. Injection of plasmid DNA carrying the coding sequences for antigen was able to decrease AHR in previously sensitized animals (Raz et al., 1994, 1996; Hsu et al., 1996). Further studies have shown that non-coding DNA sequences are important in promoting a TH1 response and this strategy is actively under investigation (Sato et al., 1996).

Non-specific Strategies. Anti-IgE , anti-cytokine, anti-receptor therapy, and anti-sense oligonucleotides provide more broad based approaches to downregulating inflammation (Chang et al., 1990; Shields et al., 1995; Renz et al., 1995; Kung et al., 1995; Kruse et al., 1992; Egan et al., 1995; Nyce and Metzger, 1997; Okudaira et al., 1996). In particular, a monoclonal anti-IgE antibody has been used in clinical trials for the treatment of allergic rhinitis (Casale et al., 1997). Treated patients showed some improvement but this was not statistically different from the placebo group. Patients did show a significant drop in free serum IgE levels and the treatment was well tolerated. It may be that a combination of a non-specific treatment such as anti-IgE antibody, plus an allergen specific treatment may be the most efficacious.

IV. CONCLUSIONS

Advances in the molecular biology of important indoor allergens including mite, cat, cockroach and fungal proteins have supported extensive research studies aimed at standardizing allergen extracts, developing new diagnostic tests, and developing new forms of immunotherapy. It is clear that recombinant allergens can be produced in large quantities and purified to homogeneity. These products show comparable immunoreactivity to the native counterparts and thus can be used for standardization of allergen extracts or for diagnosis; especially for allergens for which good extracts are not commercially available, such as *Blomia*. New forms of allergen specific immunotherapy based on recombinant allergens, and in particular peptides, have progressed to Phase 2 and 3 clinical trials, after Phase 1 trials of peptide immunotherapy for cat and ragweed allergy showed promising results. Whether this strategy can be used for allergy to complex mixtures, for example, dust mites which are a source of at least 10 allergens, needs further evaluation of T cell responses in large populations. Alternatives to peptide immunotherapy, such as allergen variants with reduced IgE Ab reactivity may be more appropriate for dust mite immunotherapy, or other multi-allergen systems. Such allergen specific treatments may be combined with non-specific approaches such as anti-IgE therapy and this type of dual therapy may prove more efficacious. It is also possible that a preventive vaccine can be formulated to promote a protective response to common indoor allergens. These new strategies require further studies, but clearly the use of recombinant allergens shows great promise for providing safer and more effective immunotherapies for asthma.

Acknowledgements: Supported by National Institutes of Health grants AI34607, AI32557, AI20565, AI38921, and a grant from the Thomas F. and Kate Miller Jeffress Memorial Trust.

REFERENCES

Casale TB, Bernstein IL, Busse WW, LaForce CF, Tinkelman DG, Stoltz RR, Dockhorn RJ, Reimann J, Su JQ, Fick RB, Adelman DC. Use of an anti-IgE humanized monoclonal antibody in ragweed-induced allergic rhinitis. *J Allergy Clin Immunol* 1997; 100:110-21.

Chang TW, Davis FM, Sun CRY, MacGlashan DW, Hamilton RG. Monoclonal antibodies specific for human IgE-producing B cells: a potential therapeutic for IgE-mediated allergic diseases. *Biotechnology* 1990;8:122-126.

Chapman MD, Smith AM, Slunt JB, Vailes LD, Arruda LK. Immunochemical and molecular methods for defining and measuring indoor allergens: in dust and air. *Pediatr Allergy Immunol* 1995;6:8-12.

Chapman MD, Smith AM, Vailes LD, Arruda LK. Recombinant Mite Allergens: New technologies for the management of patients with asthma. *Allergy* 1997;52:374-379.

Chua KY, Doyle CR, Simpson RJ, Turner KJ, Stewart GA, Thomas WR. Isolation of cDNA coding for the major mite allergen Der p II by IgE plaque immunoassay. *Int Arch Allergy Appl Immunol* 1990:91:118-123.

Chua KY, Stewart GA, Thomas WR, Simpson RJ, Dilworth RJ, Plozza TM, Turner KJ. Sequence analysis of cDNA coding for a major house dust mite allergen, Der p I. Homology with cysteine proteases. *J Exp Med* 1988:167:175-182.

Counsell CM, Bond JF, Ohman JL, Greenstein JL, Garman RD. Definition of the human T-cell epitopes of Fel d 1, the major allergen of the domestic cat. *J Allergy Clin Immunol* 1996;98:884-894.

Creticos PS, Hebert J, Philip G, and the Allervax Ragweed Study Group. Efficacy of Allervax. Ragweed peptides in the treatment of ragweed-induced allergy. *J Allergy Clin Immunol* 1997;99(no1 part 2). Abstract no.1631.

Creticos PS, Marsh DG, Proud D. Responses to ragweed-pollen nasal challenge before and after immunotherapy. *J Allergy Clin Immunol* 1989;84:197-205.

Dilworth RJ, Chua KY, Thomas WR. Sequence analysis of cDNA coding for a major house dust mite allergen, Der f I. *Clin Exp Allergy* 1991;21:25-32.

Egan RW, Athwahl D, Chou CC, Emtage S, Jehn DH, Kung TT, Mauser PJ, Murgolo NJ, Bodmer MW. Inhibition of pulmonary eosinophilia and hyperreactivity by antibodies to interleukin-5. *Int Arch Allergy Immunol* 1995;107:321-322.

Ferreira F, Hirtenlehner K, Jilek A, Godnik-Cvar J, Breiteneder H, Grimm R, Hoffman-Sommergruber K, Scheiner O, Kraft D, Breitenbach M, Rheinberger HJ, Ebner C. Dissection of immunoglobulin E and T lymphocyte reactivity of isoforms of the major birch pollen allergen Bet v1: potential use of hypoallergenic isoforms for immunotherapy. *J Exp Med* 1996;183:599-609.

Frasch CE. Haemophilus influenzae Type b conjugate and combination vaccines. *Clin Immunother* 1995;4(5):376-386.

Freeman J. Further observations on the treatment of hayfever by hypodermic inoculation of pollen vaccine. *Lancet* 1911;2:814-817.

Gavett SH, O'Hearn DJ, Li X, Huang SK, Finkelman FD, Wills-Karp M. Interleukin 12 inhibits antigen-induced airway hyperresponsiveness, inflammation, and Th2 cytokine expression in mice. *J Exp Med* 1995;182(5):1527-36.

Guerra F, Moreno C, Daza JC, Miguel R, Garcia M, Sanchez Guijo P. Linear monitoring of patients sensitive to Olea and grass pollens treated with immunotherapy based on glutaraldehyde-modified (allergoid) extracts. *J Invest Allergol Clin Immunol* 1993;3(4):182-190.

Hakkaart GA, Aalberse RC, van Ree R. Expression of the house dust mite allergen Der p 2 and mutants in the baker's yeast Saccharomyces cerevisiae. *Adv Exp Med & Biol.* 1996; 409:395-8.

Hewitt CRA, Brown AP, Hart BJ, Pritchard DI. A major house dust mite allergen disrupts the immunoglobulin E network by selectively cleaving CD23; innate protection by antiproteases. *J Exp Med* 1995;182:153-7-1544.

Higgins JA, Thorpe CJ, Hayball JD, O'Hehir RE, Lamb JR. Overlapping T cell epitopes in the group 1 allergen of Dermatophagoides spp. (house dust mite) restricted by HLA-DP and HLA-Dr class II molecules, *J Allergy Clin Immunol* 1994; 93:891-899.

Holt PG. Immunoprophylaxis of atopy: light at the end of the tunnel? *Immunology Today* 1994;15(10):484-489.

Hoyne GF, O'Hehir RE, Wraith DC, Thomas WR, Lamb JR. Inhibition of T-cell and antibody responses to house dust mite allergen by inhalation of the dominant epitope in naive and sensitized mice. *J Exp Med* 1993;178:1783-1788.

Hsu C-H, Chua KW, Tao MH, Lai YL, Wu HD, Huang SK, Hsieh KH. Immunoprophylaxis of allergen-induced immunoglobulin E synthesis and airway hyperresponsiveness in vivo by genetic immunization. *Nature Medicine* 1996;2(5):540-544.

Ishizaka K, Kishimoto T, Delespesse G, King TP. Immunogenic properties of modified antigen E. I. Presence of specific determinants for T cells in denatured antigen & polypeptide chains. *J Immunol* 1974;113: 70-77.

Julia V, Rassoulzadegan M, Glaichenhaus N. Resistance to Leishmania major induced by tolerance to a single antigen. *Science* 1996;274: 421-423.

Kruse N, Tony HP, Sebald W. Conversion of human IL-4 into a high affinity antagonist by a single amino acid replacement. *EMBO J* 1992;11:3237-3244.

Kung TT, Stelts DM, Zurcher JA, Adams GK, Egan RW, Kreutner W, Watnick AS, Jones H, Chapman RW. Involvement of IL-5 in the murine model of allergic pulmonary inflammation: prophylactic and therapeutic effect of an anti-IL-5 antibody. *Am J Respir Cell Mol Biol* 1995;13:360-365.

Lake RA, O'Hehir RE, Verhoef A, Lamb JR. CD28 mRNA decays when activiated T-cells are functionally anergized with specific peptide. *Int Immunol* 1993;5:461-466.

Lamb JR, Skidmore BJ, Green N, Chiller JM, Feldmann M. Induction of tolerance in influenza-virus immune T lymphocyte clones with synthetic peptides of influenza haemaglutinin. *J Exp Med* 1983;157:1434-1447.

Litwin A, Pesce AJ, Fisher T, Michael M, Michael JG. Regulation of the human immune response to ragweed pollen by immunotherapy. *Clin Exp Allergy* 1992;21(4):457-465.

Manetti R, Parronchi P, Giudizi MG, Piccinni MP, Maggi E, Trinchieri G, Romagnani S. Natural Killer Cell Stimulatory Factor (Interleukin 12 IL-12) Induces T Helper type 1 (Th1) specific immune responses and inhibits the development of IL-4-producing Th Cells. *J Exp Med* 1993;177:1199-1204.

Mueller GA, Smith AM, Williams DC, Hakkaart GAJ, Aalberse RC, Chapman MD, Rule GS, Benjamin DC. Expression and Secondary Structure Determination by NMR Methods of the Major House Dust Mite Allergen Der p 2. *J Biol Chem* 1997;272:26893-98.

Noon L. Prophylactic inoculation for hay fever. *Lancet* 1911;i:1572-1573.

Norman PS, Ohman JL, Long AA, Creticos RS, Gefter MA, Shaked Z, Wood RA, Eggleston PA, Hafner KB, Rao P, Lichtenstein LM, Jones NH, Nicodemus CF. Treatment of cat allergen with T-cell reactive peptides. *Am J Resp Crit Care Med* 1996;154(6):1623-1628.

Nyce JW, Metzger WJ. DNA antisense therapy for asthma in an animal model. *Nature* 1997;385:721-725.

O'Brien RM, Thomas WR, Tait BD. An immunogenic analysis of T cell reactive regions on the major allergen from the house dust mite, Der p 1, with recombinant truncated fragments. *J Allergy Clin Immunol* 1994;93:628-634.

O'Hehir R, Verhoef A, Panagiotopoilou E, Keswani S, Hayball J, Thomas WR, Lamb JR. Analysis of human T cell responses to the group II allergen of Dermatophagoides species: Localization of major antigenic sites. *J Allergy Clin Immunol* 1993;92:105-113.

O'Hehir RE, Garman RD, Greenstein JL, Lamb JR. The specificity and regulation of T-cell responsiveness to allergens. *Ann Rev Immunol* 1991;9:67-95.

Okudaira H, Mori A, Kaminuma O, Suko M. IL-5 Regulation-A New Approach to Allergy Therapy. *ACI International* 1996;8/5+6:172-179.

Raz E, Carson DA, Parker SE, Parr TB, Abai AM, Aichinger G, Gromkowski SH, Shing M, Lew D, Yankauckas MA, Baird SM, Rhodes GH. Intradermal gene immunization: the possible role of DNA uptake in the induction of cellular immunity to viruses. *Proc Natl Acad Sci USA* 1994:91d:9519-9523.

Raz E, Tighe H, Sato Y, Corr M, Dudler J, Roman M, Swain S, Spiegelberg H, Carson D. Preferential induction of a Th1 immune response and inhibition of specific IgE antibody formation by plasmid DNA immunization. *Proc Natl Acad Sci USA* 1996;93:5141-5145.

Renz H, Enssle K, Lauffer L, Kurrle R, Gelfand EW. Inhibition of allergen-induced IgE and IgG1 production by soluble IL-4 receptor (sIL-4R). *Int Arch Allergy Immunol* 1995 106:46-54.

Robinson C, Kalsheker NA, Srinivasan N, King CM, Garrod DR, Thompson PJ, Stewart GA. On the potential significance of the enzymatic activity of mite allergens to immunogenicity. Clues to structure and function revealed by molecular characterization. *Clin Exp Allergy* 1997;27:10-21.

Sato Y, Roman M, Tighe H, Lee D, Corr M, Nguyen M-d, Silverman J, Lotz M, Carson D, Raz E. Immunostimulatory DNA sequences necessary for effective intradermal gene immunization. *Science* 1996;273:352-354.

Scheiner O, Kraft D. Basic and practical aspects of recombinant allergens. *Allergy* 1995;50:384-391.

Shen HD, Chua KY, Lin WL, Hsieh KH, Thomas WR. IgE and monoclonal antibody binding by the mite allergen Der p 7. *Clin Exp Allergy* 1996; 26(3):308-15.

Shields RL, Whether WR, Zioncheck K, O'Connell L, Fendly B, Presta LG, Thomas D, Saban R, Jardieu P. Inhibition of allergic reactions with antibodies to IgE. *Int Arch Allergy Immunol* 1995;107(1-3): 308-312.

Shoji H, Hanawa M, Shibuya I, Hirai M, Yasukara T, Okumura Y, Yamakawa H. Production of recombinant mite allergen Der f 1 in insect cells and characterization of products--removal of pro-sequences is essential to IgE-binding activity. *Bioscience, Biotechnology & Biochemistry* 1996;60(4):621-625.

Smith AM, Chapman MD. Localization of antigenic sites on Der p 2 using oligonucleotide-directed mutagenesis targeted to predicted surface

residues. *Clin Exp Allergy* 1997;27:593-599.
Smith AM, Chapman MD. Reduction in IgE binding to allergen variants generated by site-directed mutagenesis: contribution of disulfide bonds to the antigenic structure of the major house dust mite allergen Der p 2. *Mol Immunol* 1996;33:399-405.
Smith AM, Chapman MD, Taketomi EA, Platts-Mills TAE, Sung SJ. Recombinant allergens for immunotherapy: A Der p 2 variant with reduced IgE reactivity retains T cell epitopes. *J Allergy Clin Immunol* 1998;101(3):423-425.
Thomas WR, Stewart GA, Simpson RJ, Chua KY, Plozza TM, Dilworth RJ, Nisbet A, Turner KJ. Cloning and expression of DNA coding for the major house dust mite allergen Der p I in Escherichia coli. *Int Arch Allergy Appl Immunol* 1988;85:127-129.
Thomas WR. Molecular analysis of house dust mite allergens. In: *Molecular Analysis of Immediate Hypersensitivity and Strategies for Immunologic Intervention.* Eds: Walker MR, Roberts AM; Wiley and Sons, Chichester UK, 1996.
Topham C, Srinivasan N, Thorpe C, Overington J, Kalsheker N. Comparative modelling of major house dust mite allergen Der p 1: structure validation using an extended environmental amino acid propensity table. *Protein Engineering* 1994;7:869-894.
Trudinger M, Chua KY, Thomas WR. cDNA encoding the major mite allergy Der f II. 1991. *Clin Exp Allergy* 21:33-37.
Van Neerven RJJ, Ebner C, Yssel H, Kapsenberg ML, Lamb JR. T cell responses to allergens: epitope specificity and clinical relevance. *Immunology Today* 1996:17(11):526-532.
Van Neerven RJJ, van de Pol MM, van Milligen FJ, Jansen HM, Aalberse RC, Kapsenberg ML. Characterization of cat dander specific T lymphocytes from atopic patients. *J Immunol* 1994;153(8):4203-4210.
Van Neerven RJJ, van't Hof W, Ringrose J, Jansen H, Aalberse R, Wierenga E, Kapsenberg M. T cell epitopes of house dust mite major allergen Der p II. *J Immunol* 1993;151:4203-4210.
Verhoef A, Higgins JA, Thorpe CJ, Marsh SG, Hayball JO, Lamb JR, O'Hehir RE. Clonal analysis of the atopic immune response to the group 2 allergen of Dermatophagoides spp.: identification of HLA-DR and DQ restricted T cell epitopes. *Int Immunol* 1993;5:1589-1597.
Wallner BP, Gefter ML. Immunotherapy with T-cell-reactive peptides derived from allergens. *Allergy* 1994; 49:302-308.
Yang X, Gieni RS, Mosmann TR, HayGlass KT. Chemically modified antigen preferentially elicits induction of Th1-like cytokine synthesis patterns *in vivo. J Exp Med* 1993;178:349-353.
Yssel H, Johnson KE, Schneider PV, Wideman J, Terr A, Kasteein R, de Vries JE. T cell activation-inducing epitopes of the house dust mite allergen Der p 1. Proliferation and lymphokine production patterns by

Der p 1-specific CD4+ T cell clones. *J Immunol* 1992;148:738-745.
Yuuki T, Okumura Y, Ando T, Yamakawa H, Suko M, Haida M, Okudaira H. Cloning and sequencing of cDNAs corresponding to mite major allergen Der f II. 1990. *Averugi-Japanese J Allergol*, 39:557-561.

Chapter 7

COCKROACH-SENSITIVE GUINEA PIG ASTHMA: A MODEL FOR INDOOR ENVIRONMENTAL ASTHMA

Bann C. Kang, M.D., Daohong Zhou, M.D., Gang Chen, M.D., Robyn Proffitt, B.S., Beth Miller, M.D., and Jin Tae Kim, M.D.

I. OVERVIEW

Asthma is a worldwide disease with significant morbidity and mortality. Despite considerable progress in asthma research, we still have much to learn about its genetics, etiology, and pathogenesis. Particularly, the rising mortality among high risk asthma patients in the inner-city population has been a major concern in recent years (Panlozzi et al., 1984; Evans et al., 1987; Weiss, 1990). Over 85% of the inner-city asthma population have atopic asthma and are allergic to a variety of inhalant allergens (Hendrix et al., 1975; Ford, 1983; Kang et al., 1993); a significant proportion (22 to 79%) of this population is also found to possess cockroach hypersensitivity (Benton & Brown, 1967; Mendoza & Snyder, 1970; Lan et al., 1988; Kang et al., 1989; Sastre et al., 1996). The significance of cockroach hypersensitivity in asthmatic subjects has been recognized after it was learned that cockroach allergen (CRa) inhalation induces early asthmatic reaction (EAR) as well as late asthmatic response (LAR) in cockroach-sensitive asthmatic subjects (cockroach asthma) (Kang et al., 1979; Kang et al., 1992). Moreover, it has been found that these subjects show a higher rate of dual asthmatic responses than do asthmatic subjects who are sensitive to other inhalant allergens such as ragweed (Kang et al., 1992). In addition, it has been found that asthmatic subjects with cockroach sensitivity commonly show more severe and frequent exacerbations of asthma, because of which the subjects heavily utilize emergency room services and are more prone to hospitalization (Mendoza & Snyder, 1970; Kang et al., 1979; Kang et al., 1989; Kang et al., 1993). Recently, it was reported that the severity of asthma in inner-city asthmatic children is closely associated with the cockroach hypersensitivity of patients who live in inner-city dwellings heavily infested with cockroaches. Hypersensitivity to cockroach correlates with the cockroach allergen (Bla g 1) levels in the house dust of the homes in which asthmatic children of urban minorities of low socioeconomic status tend to live (Sarpong et al., 1996; Rosenstreich et al., 1997). As a matter of fact, the cockroach allergen content has been confirmed to be the highest of all known indoor allergens in the dust of dwellings of inner-city asthmatic children of several major cities in the states.

The relevance of a confined environment for the spread of infectious disease in humans has been well established. It also has been shown that

1-56670-373-5/99/$0.00+$.50

toxins or antigens in poorly ventilated work places induce occupational lung diseases. However, only recently has the importance of allergens in the indoor environment, particularly the allergens of the home environment, been recognized for affecting humans with asthma. The quantity as well as the content of the indoor environment appears to affect the sensitization of the hosts (Pollart et al., 1989; Sporik et al., 1990; Platts-Mills et al., 1991). Particularly important is the quantity, or threshold levels, of certain allergens that can be airborne and function as aeroallergens, such as cockroaches, dust mite, pet allergens, and mold spores. Exposures to cockroach allergen occur in children at home, as well as at school, when cockroach allergens in the dust become airborne after rooms are disturbed by children's activity (Gelber et al., 1993; Rosenstreich et al., 1997; Sarpong et al., 1997). The significance of the cockroach allergens in the indoor environment, however, lies in the numbers of asthmatic subjects involved as well as in the severity of asthma in those who are afflicted with the cockroach asthma. Cockroach infestation is a worldwide phenomenon and it does not discriminate between urban and suburban locations or geographical locations (Roth & Willis, 1957; Koehler et al., 1987). Yet, sensitivity is much higher in northern cities than in southern or west coast cities (Call et al., 1992; Sperber et al., 1993; Christiansen et al., 1996). The difference might be related to the amount of time the hosts spend in a confined environment as well as to the level of the allergens generated by home ventilation and climate controls.

Avoidance of allergen has been reasonably successful in managing the exacerbations of acute asthma and preventing further sensitization for a few specific allergens such as dust mites and pet allergen(s) (Pollart et al., 1989; Sporik et al., 1990; Platts-Mills et al.,1991, Gelber et al., 1993). However, complete avoidance of the infested insect from the confined environment is a formidable task, as the insects cannot be easily eradicated by the usual pest control technology. Consequently, resolving problems of cockroach sensitivity and cockroach asthma become difficult, if not impossible, tasks. But because cockroach-sensitive asthma constitutes a major portion of inner-city asthma, tackling this asthma is necessary and it would contribute to reducing the morbidity and mortality of inner-city asthma as a whole. An elucidation of the pathogenesis of cockroach-sensitive asthma will likely lead us to a better understanding and a better management of the inner-city asthmatic population. Furthermore, the knowledge gained in the study of cockroach-sensitive asthma can be extended to the pathogenesis of other indoor allergen-induced allergic asthma in general. Therefore, we have developed a cockroach-sensitive animal asthma model to study the mechanism and pathogenesis of cockroach-sensitive asthma.

The use of animal models to study human diseases such as asthma has been a practical tool despite obvious shortcomings. An ideal asthma model would be one in which the asthma created in a given animal would possess almost all the characteristics of human asthma, in this case, cockroach asthma in humans. Various animal models have been used to study bronchial asthma; the guinea pig model has been one of the most frequently used animal models (Perini & Mota, 1973; Anderson, 1980; Karol, 1994; Santing et al., 1994; Andersson et al., 1995). Guinea pig asthma is known to resemble human asthma in many ways, including its distinct similarities in asthmatic responses *in vivo* as well as airway muscle contractilities (Shellenberg, 1995). However, unlike other animals, the guinea pig develops two types of anaphylactic antibodies, IgE and IgG_1, when it is sensitized (Perini & Mota, 1973; Karol, 1994). The mode of immunization and the route of antigen entry influence the development of the antibodies (Kang et al., 1995). Antibody development might also be influenced by different allergens (ovalbumin, TDI, and dust mites, etc) along with or without the use of an adjuvant. Yet, a great majority of studies using these animal asthma models has not identified the type of reagins (Anderson, 1980; Santing et al., 1994; Andersson et al., 1995; Shellenberg, 1995). A commonly used immunization method consists of a few intraperitoneal or intradermal injections of the allergen with pertussis vaccine or $Al(OH)_3$ as an adjuvant (Perini & Mota, 1973; Anderson, 1980; Andersson et al., 1995; Mapp et al., 1996). Some investigators have achieved allergen sensitization by aerosol or nasal instillation plus initial intraperitoneal injection as a booster (Ishii et al., 1989). However, we have adopted a method that uses only chronic aerosolized exposures (Kang et al., 1996). This approach more closely simulates allergen exposure in a home environment. Therefore, guinea pigs were chronically immunized in a steady manner repeatedly without using any adjuvant.

II. THE CHARACTERISTICS OF COCKROACH-ALLERGIC ASTHMA IN HUMANS

Identification of relevant allergens in asthmatic subjects is usually determined by documentation of the presence of the specific anaphylactic antibody, so-called reagin(s), using either allergy skin test *in vivo* and/or by the detection of the antigen-specific IgE antibody-RAST. Therefore, the allergic status of an asthmatic subject theoretically will not be known until the subject has been investigated with the appropriate antigens. A high index of suspicion followed by the appropriate investigation with proper allergens is a prerequisite in determining the sensitivity for a given asthmatic subject. Recently, research by allergy-oriented investigators has identified more and more allergens that cause asthma; the result has been that some cases of idiopathic asthma with unknown cause have been determined to be allergic asthma with newly discovered causative agent(s).

The cockroach-allergic asthma is one of such newly recognized allergic entities. The cardinal features of allergic (cockroach) asthma are reversible airflow obstructions, early asthmatic response (EAR) that occurs a few minutes to an hour after the antigen inhalation, and/or late asthmatic response (LAR) that starts in a couple of hours after allergenic inhalation as a slow and sustained airway obstruction lasting from a few hours to 24 hours. Moreover, a great majority of subjects with allergic asthma develop both EAR and LAR, called dual asthmatic responses (Patterson et al., 1995; Seth & Lemanske, 1995). Some allergic asthmatic subjects may show a slow persistent airway obstruction in response to a single inhalation challenge (Perrin et al., 1991; Kang et al., 1992). At any rate, such obstructive airway responses as seen in allergic asthmatic subjects are believed to be due to underlying chronic airway inflammation created by mast cell- mediated allergic reactions in the airways. Evidence of allergic airway inflammation has been documented directly by the presence of inflammatory cells in bronchoalveolar lavage fluids (BALF) and/or by the direct histopathological examination of the lung tissues. Moreover, airway inflammation is further supported by increases in inflammatory mediators in the airways as well as by the documentation of cytokine production in the airway cells (Hallsworth et al., 1994; Ying et al., 1995; Humbert et al., 1997). Airway inflammation has been proven to exist in most asthmatics, and is not limited to allergic asthma.

In subjects with cockroach-sensitive asthma, i.e., those defined as being asthmatic with cockroach hypersensitivity as demonstrated by a positive skin test to cockroach antigen, a great majority (87%) give a positive inhalation challenge test (Kang et al., 1979; Kang et al., 1992). Cockroach hypersensitivity can exist as an isolated allergen sensitivity or as a part of multiple inhalant hypersensitivity. However, not all cockroach-sensitive subjects have asthma. Most of the subjects with cockroach asthma are known to have had persistent symptoms of chronic duration. Most, but not all subjects with cockroach asthma also have elevated specific IgE antibodies, although not all have an elevated total IgE antibody level. Cockroach antigen challenge in the subjects with cockroach asthma provokes variable degrees of airway obstructions at different time points after the antigen challenge, i.e., EAR, LAR, or dual or persistent patterns. Yet, it is more common to see dual or persistent airflow obstruction than simple EAR in these subjects (Kang et al., 1979; Kang et al., 1992; Kang et al., 1993). Although cockroach asthma is usually found among inner-city minorities of lower socioeconomic status, it is not uncommon to find it among professionals or laboratory workers who handle the insects (Steinberg et al., 1987; Kang et al., 1989). Moreover, it has recently been documented that the sensitivity to cockroach allergen and the severity of asthma among inner-city children are highly correlated with the level of CRa (Blag 1) antigen in the house dust, particularly the dust in the bedroom (Rosenstreich et al., 1997). However, only a minority of subjects who dwell in insect-infested dwellings develop cockroach-sensitive asthma, suggesting that other factors are involved in the development of cockroach asthma, e.g., genetic and personal risk factors, etc., in addition to allergens in the house.

III. DEVELOPMENT OF AN ANIMAL MODEL TO STUDY COCKROACH ASTHMA

A. SELECTION OF ANIMAL

After choosing the guinea pig as the animal to sensitize for the reasons stated in the overview, we used the English Shorthair strain for initial study, and later we changed to Hartley strain as they show asthmatic responses more consistently in a conscious state. We have also compared cockroach allergens (CRa) sensitization on an inbred strain, NCI 21, to determine whether there is any variation in strains. We discovered that there is a certain strain difference in the level of airway responsiveness *in vivo*, although all three strains show reaginic sensitization by both skin test and PCA after chronic aerosolized exposures (Kang et al., 1997). Eighty-three percent of CRa-sensitized Hartley guinea pigs show some form of airway obstructions *in vivo,* while 50% of the sensitized NCI 21 and English Short hair strains reveal signs of airway obstruction in response to CRa challenges. Therefore, Hartley guinea pigs in a conscious state were mainly used in our cockroach asthma studies. However, there was consistently significant airway obstruction in English Short hair strains when they were studied under anesthesia (Kang et al., 1996).

B. METHOD OF ALLERGEN SENSITIZATION

For the reasons stated in the overview, CRa was chosen as an indoor allergen to create a cockroach-sensitive animal model closely resembling cockroach asthma in humans. Crude German cockroach extract instead of a well characterized cockroach allergen fraction, e.g., Blag 1, was used for the same reason. Humans who live in infested dwellings are most likely exposed to allergens that contain all the components of disintegrated materials from the whole body of the insects. (Kang et al., 1989). German cockroaches are the most common infestants of houses and apartments in the northern inner-cities of the U.S. (Bernton & Brown 1967; Koehler et al., 1987). In a preliminary study we tested for an optimal method for CRa sensitization in guinea pigs (Kang et al., 1995). Based on the methods of Bottomley et al., six different methods of aerosolized CRa exposure were tried. The aerosolized sensitizations were compared with a method of CRa sensitization by intraperitoneal injection without an adjuvant (see Table 1). The CRa sensitization was determined by measuring IgG_1 and IgE antibodies produced at various time points of the sensitization using passive cutaneous anaphylaxis (PCA). Table 1 shows that anaphylactic antibodies developed when guinea pigs were exposed to aerosolized CRa in relatively high doses. Repetitive high dose sensitization was more consistently effective in producing IgG_1 antibody. However, CRa injections by the intraperitoneal route produced both IgE and IgG_1 antibodies (Kang et al., 1995). More interestingly, however, we have found that the guinea pigs sensitized by aerosolized CRa developed only IgG_1 antibody and responded to CRa challenge with airflow obstruction *in vivo*, even though these animals had no IgE antibody. Also, we have discovered that aerosolized CRa did not act as an adjuvant to a known allergen, ovalbumin (OA), with sensitization by aerosols. The production of CRa-specific IgG_1 in the CRa aerosol-sensitized guinea pigs is not related to characteristics of the CRa itself as commonly used antigens, such as dust mite or TDI, also induced antigen-specific IgG_1 antibody but not IgE (Kang et al., 1995; Mapp et al., 1996, Hsiu et al., 1997).

C. ALLERGEN SENSITIZATION

A lyophilized whole body extract of German cockroaches was obtained from Greer Laboratories, Inc. (Lenoir, NC) and used as the allergen in sensitization and in challenges after it was reconstituted appropriately.

Hartley strain guinea pigs, male, weighing 200-250g, were purchased from Hilltop Laboratories. Upon arrival, the animals were housed in a plastic cage in a colony room with a 12 h day-night cycle (lights on at 0700h) achieved through artificial illumination. They received food and water *ad libitum*. Prior to the start of the experiment guinea pigs were quarantined and acclimatized to their housing for one week. The experiment was carried out with the approval of the Institutional Animal Care and Use Committee at the University of Kentucky. CRa aerosol sensitization was varied slightly for different studies, but mostly stayed with a constant dose of CRa aerosol exposures repeated twice daily, 5 days per week for 4 weeks as described (Kang et al., 1996). Briefly, guinea pigs were placed in a semi-closed acrylic chamber connected to an ultrasonic nebulizer (Devilbiss Model 100, Somerset, PA) generating aerosol; droplet sizes ranged from 0.5 to 30 um, with an output of 0.2 ml/min. Animals were exposed to 0.5 mg-25 mg CRa in 10 ml saline, for 45 min, twice a day, 5 days a week, for a total of 4 weeks. In a study of the dose-responsiveness in CRa sensitization, we used three concentrations of CRa: 0.5 mg (low dose, LD), 5 mg (medium dose, MD), or 25 mg (high dose, HD). A group of guinea pigs (Control) were sensitized with aerosolized saline (vehicle) in the same manner as used for CRa sensitization.

D. DETECTION OF ANAPHYLACTIC ANTIBODY

1. Allergic Skin Test (AST)

During sensitization of guinea pigs the level of the sensitization was tested on days 19 and 29/30 by AST. The skin test was performed after the animals were lightly sedated with sodium pentobarbital (30 mg/kg) and the back skin was shaved. Using 27G test syringes for intradermal injection, 27g X 3/8 inch syringes (Terumo Medical Co., Elkton, MD) were used to deliver a series of CRa (0.5mg , 1mg and 5mg) injections in 0.03 ml of saline with positive (30 mg) and negative controls into the upper dermis intradermally. Two diameters, lengthwise and vertical, of a "wheal" were measured at 20 and 60 minutes respectively after the inoculations. A mean of two diameters was used.

Table 1. Protocols of CRa sensitization and their results in antibody production

Groups	Regimen of sensitization	Blood drawn in days after immunization	PCA Titers		
			$IgG_1(4h)^a$	IgE (8 days)[b] nonheat	heat
Aerosol:					
Low Dose					
C-I	0.5mg daily for 20 days	19,34,49	0	0	ND
C-II	0.5mg, 5 days/wk for 4 wks	19,34,49	0	0	ND
C-III	5mg (0), 0.5mg (5), 0.005mg (19)	19,34,49	0	0	ND
High Dose					
C-IV	50mg (0), 5mg (5), 0.05mg (19)	19,34,49	0	0	ND
C-V	5mg 5 days/wk for 4 wks	22	1:2 (8X9)	0	ND
		37	1:2 (9X9)	0	ND
		52	1:20 (10X11)	0	ND
C-VI	50mg every other day for 1 wk, plus 5mg every other for 5 wks	19	1:20 (13X14)	0	ND
		34	1:200 (13X16)	0	ND
		49	1:20 (11X13)	0	ND
Intraperitoneal:					
C-VII	50mg (0), 5mg (5), 0.05mg (19)	19	1:20 (6X7)	1:2 (13X14)	0
		34	1:20 (9X14)	1:1 (18X19)	0
		49	1:20 (9X11)	1:2 (11X15)	0

ND: Not done a: latent period of 4 hours b: latent period of 8 days

2. Passive Cutaneous Anaphylaxis (PCA) Assay

In an earlier study, PCA assays were done with guinea pig blood drawn on days 19, 34, and 49 in order to trace the kinetics of anaphylactic antibody production. In later studies, PCA was done after completion of the sensitization, ie. in week 4. Both cytophilic antibodies, IgG_1-like antibody (IgG_1) and IgE-like antibody (IgE), of the sensitized guinea pigs were measured by PCA reactions as previously described (Kang et al., 1995; Kang et al., 1996). Pooled sera were divided into two aliquots, one for IgG_1 and the other for IgE. For IgG_1, the PCA assay was done at a latent period of 4 hours after sera were inoculated. Sera for IgE antibodies were also divided into two aliquots, one nonheated and the other heated (56° C, 30 min); both sera were tested for PCA after 8 days of the latent period. Intradermal injections of 0.1 ml of serially diluted sera (1:2 to 1:2,000) in duplicate were injected intradermally at 12-16 sites into the shaved back skin of the guinea pigs (English Shorthair). After an appropriate incubation period, the relevant antigen (2 mg CRa or OA in some cases) was mixed with Evans blue, 10 mg in 1 ml of isotonic-buffered saline (0.1 *M* NaCl), and was infused into an ear vein of the animal. Bluing reactions were recorded 30 min after the antigen infusion. The highest titer was judged from the dilution of the sera showing a positive PCA reaction in duplicate, i.e., bluing diameters over 5 mm. Sera collected from the guinea pigs of each group (LD, MD, HD, and Control) were pooled by the group and used for PCA assays

E. MEASUREMENT OF LUNG FUNCTION CHANGES FOLLOWING CRa CHALLENGE

1. In Anesthetized Guinea Pigs

An appropriately immunized guinea pig was anesthetized with an intraperitoneal injection of sodium pentobarbital (30-35mg/kg), cannulated for blood drawing, intubated with an endotracheal tube, paralyzed with Gallamine triethiodide administered intravenously (10mg/kg), and artificially ventilated. The animal was placed supine inside a 2.8-liter whole-body plethysmograph. The maximal expiratory flow volume maneuver was carried out as described previously (Lai, 1988; Lai & Zhou, 1997). Briefly, the flow rate was monitored with a Statham PM 15 differential pressure transducer as the pressure dropped across three layers of 325-mesh wire screen in the wall of the plethysmograph. The lung volume change was measured via integration of flow. Airway opening pressure (Pao) was measured and the lung was inflated 3 times to total capacity (lung volume at Pao=30 cm H_2O). At peak volume during the third inflation, the inflation valve was shut off, and automatically another solenoid valve for deflation was turned on. The deflation valve was connected to a 20-liter container with a negative pressure of 40 cm H_2O. This negative pressure produced the maximal expiratory flow (Vmax). The maximal expiratory flow volume plot was recorded and stored on a cathode ray storage oscilloscope (Hitachi, V-134). Respiratory compliance (Crs) was obtained by dividing the tidal volume by its accompanying Pao difference during artificial ventilation. The Pao difference was measured between end inspiration and end expiration (i.e., instrument zero flow). The maximal expiratory flow rate at 50% total lung capacity (Vmax50) was obtained.

After baseline measurements of lung function were made and blood for leukotrienes (1.0cc) was drawn via a carotid artery cannula, the animal was challenged with CRa, 1mg/10ml of buffered saline. The CRa was aerosolized by an ultrasonic nebulizer and delivered via tubes connected to the respirator for a duration of 30 min. The Harvard respirator used was set at a rate of 60 breaths per minute with a tidal volume of 8-10 ml/kg.

2. In Conscious Guinea Pigs

Asthmatic responses of the CRa-sensitized guinea pigs in a conscious state were studied as described (Miller et al., 1996; Zhou et al., 1998); upon completion of CRa or control immunization guinea pigs were allowed to rest for at least 5 days in their home cages; then the animals were tested for CRa provocation *in vivo*. The day before the CRa provocation, guinea pigs were acclimated to the double-chamber plethysmograph (Buxco Electronics, Inc., Troy, NY) for at least 4 hours to minimize the influence of restraint stress. On the day of the CRa provocation experiment, guinea pigs were placed into the chamber and allowed to settle down while monitoring of specific airway resistance (SRaw) was continued with the guinea pigs in a conscious state. A segment of 10 minutes of data is used as baseline SRaw. The guinea pigs were then challenged with a nebulized CRa aerosol (1mg/ml) at a flow rate of 3 liters/min controlled by a bias flow machine for 10 min and for 30 min. for the second challenge. The interval between the two challenges was 10 minutes. SRaw was recorded immediately after the first challenge and then continuously monitored. Data were collected as a mean value for 10 minutes of every hour up to 8 hours. The guinea pigs were then released and returned to their home cages for resting and feeding overnight. Then, SRaw was again measured the next day at 24 hours post-challenge. The method of lung function measurement was modified from the method described by Sugiyama, et al. (1992).

F. INFLAMMATORY CELL ANALYSIS IN BRONCHOALVEOLAR LAVAGE FLUID (BALF)

Immediately following the 24 hour measurement of lung function in a conscious state, the guinea pigs were anesthetized by an injection of sodium pentobarbital (30 mg/kg), intraperitoneally. The trachea was exposed and cannulated. Bronchoalveolar lavage was conducted by instillation of sterile saline (5ml/time, two times) through the tracheal cannula; BALF was collected (about 70% volume of instilled solution) and analyzed for cell counts (Sugiyama et al., 1992). The collected BALF was centrifuged at 1500 rpm for 10 minutes. The pelleted cells were treated with Tris-buffered ammonium chloride solution (pH 7.2) to lyse red blood cells. The remaining cells were washed once with PBS supplemented with 1% FCS and 0.1% antibiotics (from Gibco, Long Island, NY). Total cell counts were determined using a hemocytometer. Differential leukocyte counts were undertaken on cytospin slides, which were prepared by centrifuge of BALF cells on a glass slide using a Cytospin centrifuge and stained with a Diff-Quick staining kit (Jorgensen Labs., Inc., Loveland, CO). A minimum of 200 leukocytes were counted using standard morphologic criteria .

G. MEASUREMENT OF AIRWAY SMOOTH MUSCLE CONTRACTILITY *IN VITRO*

Study of contractile responses of airway smooth muscles *in vitro* was done as described previously (Southrada et al., 1992; Chen et al., 1997). After completion of BAL, the guinea pigs were euthanized, and the trachea was isolated and soaked in ice-cold Krebs solution (NaCl, 118mM; KCl, 4.8mM; $CaCl_2$, 2.5mM; $MgSO_4$,1.2mM; $NaHCO_3$, 24.9mM; KH_2PO_4, 1.15mM; glucose, 5.6mM; and HEPEP, 12.6mM, pH 7.4) with aeration by 95%O_2-5%CO_2. After clearing excess fat and connective tissues, the trachea was cut into 3-mm rings and mounted in organ baths containing Krebs solution (37° C). The tracheal rings were initially stretched to a resting tension of 0.75g and equilibrated for 60-90 min with replacement of buffer every 20 min. Isometric tension development was measured using force-displacement transducers (Model FT-03C, from Grass Instrument Co., Quincy, MA) and an EEG and polygraph data recording system (Model 79D, from Grass Instrument Co.).

H. MEASUREMENT OF LEUKOTRIENES B_4 AND C_4/D_4 IN ARTERIAL BLOOD

Leukotrienes B_4 (LTB_4) and C_4/D_4 (LTC_4/D_4) were measured using an ELISA Technologies kit (product B_4: No. 406110 and C_4/D_4: No. 406210). Reagents for leukotrienes, LTB_4, and LTC_4/D_4, were bought from ELISA Technologies (Lexington, KY). Heparinized blood was drawn from a precannulated carotid artery at 0, 5, 45, and 360 min after completion of the CRa challenge and stored at -20° C for later use. The minimal detectable level was 0.02 ng/ml with a constant variance of 10%.

IV. CHARACTERISTICS OF THE COCKROACH-SENSITIVE GUINEA PIG

During the last several years we have conducted a series of investigations using the cockroach allergen-immunized guinea pig and have established cockroach-sensitive guinea pigs with asthma, a model whose asthma very closely resembles the human asthma noted in asthmatic subjects with cockroach sensitivity.

A. PRESENCE OF ANAPHYLACTIC ANTIBODY

In the earlier studies we demonstrated that repeated exposures of CRa aerosol alone can sensitize guinea pigs without the help of an adjuvant. Two of 6 methods using the aerosol route induced the guinea pigs to produce a reaginic antibody at day 19 by PCA. The reaginic antibody was IgG_1, not IgE. However, guinea pigs immunized with CRa via the intraperitoneal route produced both types of reaginic antibodies, IgG_1 and IgE, which was also achieved without the help of an adjuvant. The IgG_1 antibody that was produced remained detectable up to day 52, even though the exposure was terminated at day 28. It appears that CRa sensitization by aerosolization requires either frequent exposures of a certain threshold or high doses, either repeatedly or intermittently. The production of IgG_1 antibody appears not to be influenced by co-inhalation of AL (OH_3), an adjuvant, although it is customary to use such an adjuvant such as either pertussis vaccine or Allum hydroxide gels during animal sensitization (Perini & Mota, 1973; Anderson, 1980; Karol, 1994; Santing et al., 1994; Bramley et al., 1995). The CRa aerosol immunization, however, produces antibody in generally lower titers, 1:2 and up to 1:200 dilutions, than does intraperitoneal immunization (Mapp et al., 1996). Later studies confirmed that the sensitized state (presence of anaphylactic antibody) was also easily identifiable by skin test *in vivo* (Kang et al., 1996; Zhou et al., 1998). The absence of IgE production does not seem to influence the skin test at least in this guinea pig system. Similar studies in the production of low titer of IgG_1 antibody have been reported by other investigators, including a recent paper by Hsiu, et al., (1997). The levels of anaphylactic antibody (by skin test) produced by the repeated CRa aerosol exposures were loosely correlated to the amount of CRa given during the immunization period (Zhou et al., 1997; Zhou et al, 1998) (see Table 2).

Table 2. Results of anaphylactic antibodies produced by CRa sensitization in guinea pigs

Group	Allergy Skin Test			PCA Titers		
	Saline	Histamine	CRa 1 ug	4 Hours (IgG1)	7 Days (IgE) Non-heat	7 Days (IgE) Heat
Control (n=11)	0.0 ± 0.0	12.3 ± 0.4	0.0 ± 0.0	0 (0X0)	0 (0X0)	0 (0X0)
LD (n=6)	0.0 ± 0.0	14.3 ± 0.3	9.2 ± 0.7	1:2 (6X6)	0 (0X0)	0 (0X0)
MD (n=6)	0.0 ± 0.0	12.8 ± 0.3	10.0 ± 0.6	1:40 (9X8)	0 (0X0)	0 (0X0)
HD(n=6)	0.0 ± 0.0	13.4 ± 0.1	11.0 ± 0.2	1:160 (9X7)	1:5 (9X5)	1:5 (8X6)

Modified from Zhou et al. 1998.

The skin test results are presented as means ± SE, size of "wheals", at 60 min post-saline, histamine, or CRa injections on day 29. The PCA was done on the sera pooled from the same groups collected on day 35 and the results are expressed as the titers. Values in parentheses are the mean diameters of the blue area measured at 30 min post CRa challenge in duplicate. Control: Saline-sensitized, LD: Low dose, MD: Medium dose, and HD: High dose, of CRa-immunized guinea pigs.

B. PRESENCE OF AIRWAY OBSTRUCTION IN RESPONSE TO THE ANTIGEN CHALLENGE

The guinea pigs sensitized by exposure to CRa aerosols revealed asthmatic responses in response to CRa challenges. Airway obstruction was noted in CRa-immunized guinea pigs both under anesthesia and in conscious states.

In a preliminary study the obstructive airway responses were apparent in the CRa-immunized animals immediately upon CRa inhalation at one and 4 hours after challenge in the anesthetized guinea pigs (Kang et al., 1995). Ventilatory function, Vmax50 and respiratory compliance (Crs), were reduced to 38% and 62% of the respective baseline values, while the other guinea pigs showed severe airway obstruction, 57% and 56%, respectively at 4 hours post-challenge. By contrast, the sham-immunized guinea pigs (three animals) showed little change in either Vmax50 or Crs throughout the 4-hour period. Moreover, lung pathology of the CRa-immunized guinea pigs revealed an increase in cellularity with abundant infiltration of eosinophils. Such an apparent antigen-specific airway obstruction was further confirmed in a series of studies done *in vivo,* both under anesthesia (Kang et al., 1996) and in conscious animals (Miller et al., 1996; Zhou et al., 1997; Zhou et al., 1998). Again, the signs of airway obstruction, eg., reduced ventilatory function depicted in the Vmax50 as well as in Crs in the anesthetized and intubated guinea pigs, were apparent only in the CRa-immunized groups. Moreover, the CRa-specific airway obstructions lasting up to 6 hours, noted in the anesthetized guinea pigs correlated with elevated leukotrienes, LTB_4 and LTC_4/D_4, in arterial plasma post-pulmonary circulation (Kang et al., 1996).

We further confirmed the presence of airway obstruction in response to cockroach allergen inhalation (asthma-like response) in the CRa-immunized guinea pigs in a conscious state. In order to remove any unwanted effect on the ventilatory function of the guinea pigs, possibly caused by anesthetic agent(s), we repeated the CRa-challenge test and observed the conscious, unmedicated guinea pigs for a longer period. Two groups of Hartley guinea pigs were immunized with CRa or saline (Sham) using the same method and challenged with CRa aerosol at day 35/36.

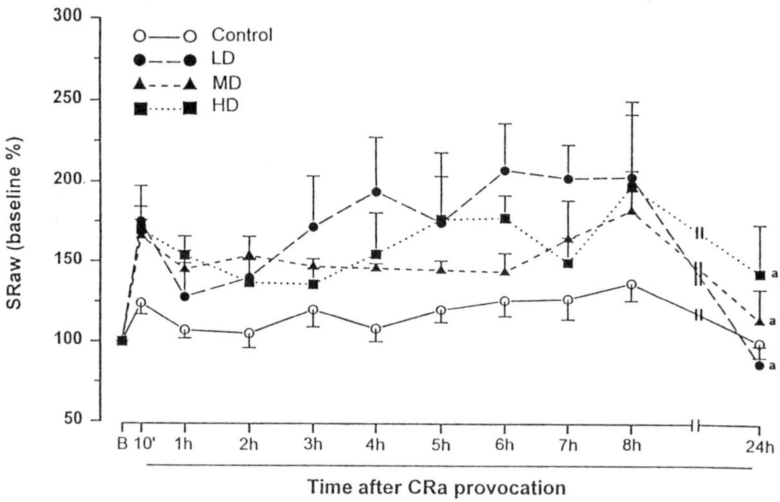

Figure 1. Changes in specific airway resistance (SRaw) following CRa-provocation. *The SRaw data taken after CRa challenge are expressed as percent of baseline, mean ± SE. Control: Saline-immunized (n=11); LD: Low dose-(n=6); MD: Medium dose-(n=6); HD: High dose-(n=6) of CRa-immunized guinea pigs. a: p<0.01 CRa-immunized vs. Control.*

Changes in ventilatory function and airway resistance (SRaw) were recorded continuously using a Buxco double-chamber plethysmograph. Although individual guinea pigs showed an increase in SRaw at various time points, the average changes in SRaw revealed the EAR and LAR rather clearly. Increases in SRaw (from baseline), EAR in one hour, and LAR between 3 hours and 8 hours after the challenge. There were also marked variabilities in the pattern of airway obstructions among individual animals, e.g., from guinea pigs not responding at all to CRa to dual responders, isolated early responders, or isolated late responders. A few sensitized guinea pigs responded to CRa challenge with a persistent airway obstruction throughout the observation period. Overall, CRa challenge induced EAR and LAR in the CRa-immunized guinea pigs. Fig. 1 shows the changes in SRaw in three groups (LD, MD, and HD) of the CRa-immunized guinea pigs as compared to sham-immunized (control) guinea pigs. Interestingly, however, this figure also reveals that there was not a clearcut dose-responsiveness in the CRa-induced asthmatic responses in conscious animals, i.e., animals receiving high amounts of CRa (HD group) during sensitization period did not show a proportionately greater increase in SRaw than those of the LD or MD groups. This was, however, contrary to the ample evidence found that there was a dose-responsiveness in reactivity such as, PCA, leucocytosis with eosinophilia in BALF, and airway smooth muscle contractilities upon CRa challenges.

C. AIRWAY SMOOTH MUSCLE CONTRACTILITY *IN VITRO*

In order to elucidate the mechanism of the CRa-induced airway obstruction in immunized guinea pigs, the contractile responses of the airway smooth muscles of the immunized guinea pigs were examined *in vitro*. Airway smooth muscles of the appropriately immunized guinea pigs were isolated at 24 hours post-antigen challenge and were tested for their contractile responses to non-specific stimuli, such as electric field stimulation (EFS) and acetylcholine (Ach), as well as to the specific allergen, CRa, *in vitro* (Chen et al., 1998).

The airway muscles of the CRa-immunized guinea pigs responded with CRa-specific contractions in a dose-dependent manner. Fig. 2A shows the comparative contractile sensitivities to a series of four doses of CRa challenges to the tracheal smooth muscles isolated from three groups of the CRa-immunized guinea pigs (LD and HD) as compared to those of control-immunized guinea pigs. The data reveal that the CRa-specific contractile force varies by the amount of CRa to which animals were exposed during the immunization period. In fact the contractile sensitivity directly correlates with the CRa doses at least in this experiment ($p<0.001$). However, the contractile responses of the immunized guinea pigs to non-specific cholinergic stimulants, both EFS and Ach, were essentially the same as those of control-immunized guinea pigs (Fig 2B).

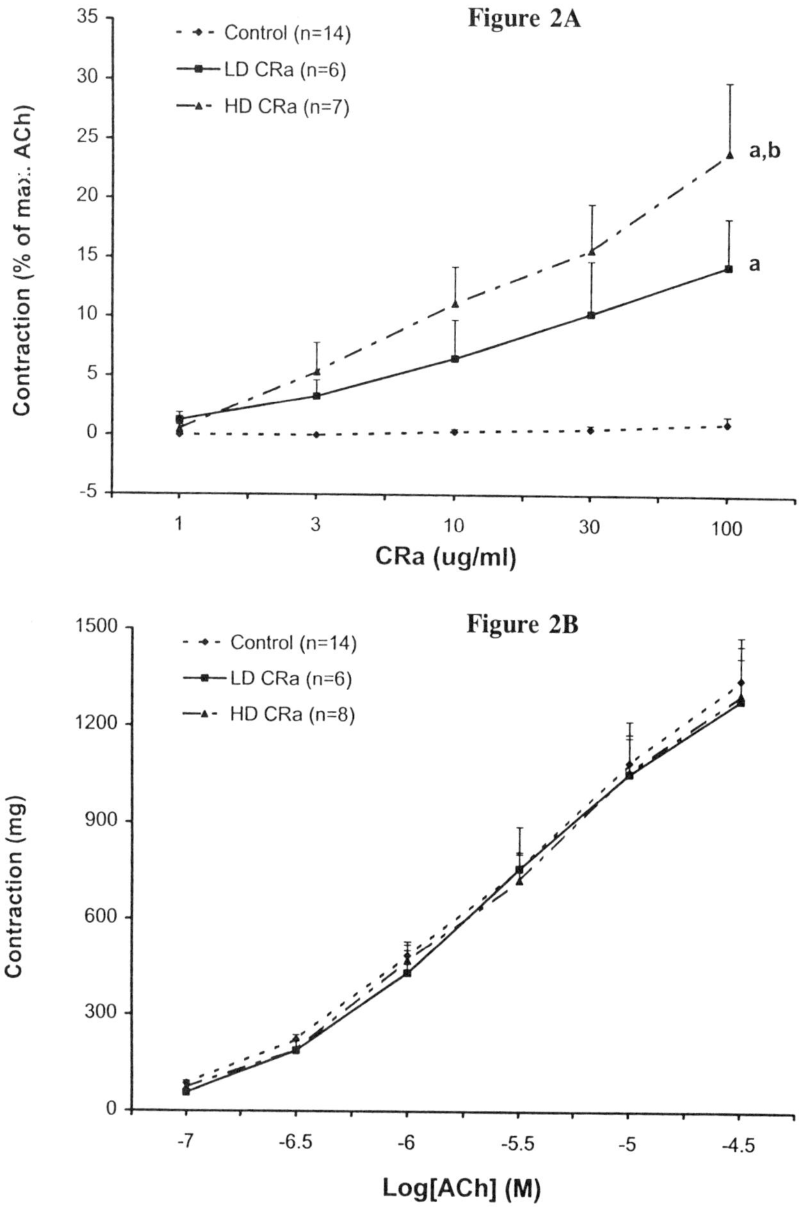

Figure 2 - A&B. Cumulative dose-response curves to CRa- (A) and Ach-stimulations (B). *The tracheal smooth muscle rings, isolated from saline- (Control), low dose (LD), or high dose (HD) of CRa- immunized guinea pigs at 24 hours post CRa inhalation challenge, were stimulated with cumulative doses of CRa (A) and Ach (B). Data are expressed as percentages of maximal contraction to ACh ($3X10^{-5}$ M) stimulation (A)and isometric tension in mg (B), mean ± SE. a: $p<0.001$ vs. control, b: $p<0.001$ vs. LD.*

The amount of CRa with which guinea pigs were immunized, e.g. LD, MD, or HD, did not alter the sensitivity of the immunized airway smooth muscles to cholinergic stimulants. In addition, the immunized airway muscle rings did not contract in response to a non-specific allergen, OA, at all. Moreover, there was also little qualitative difference in the contractile responses between tracheal and bronchial airways, except that bronchial smooth muscles showed slightly increased sensitivity compared with the tracheal airway rings of the same guinea pigs.

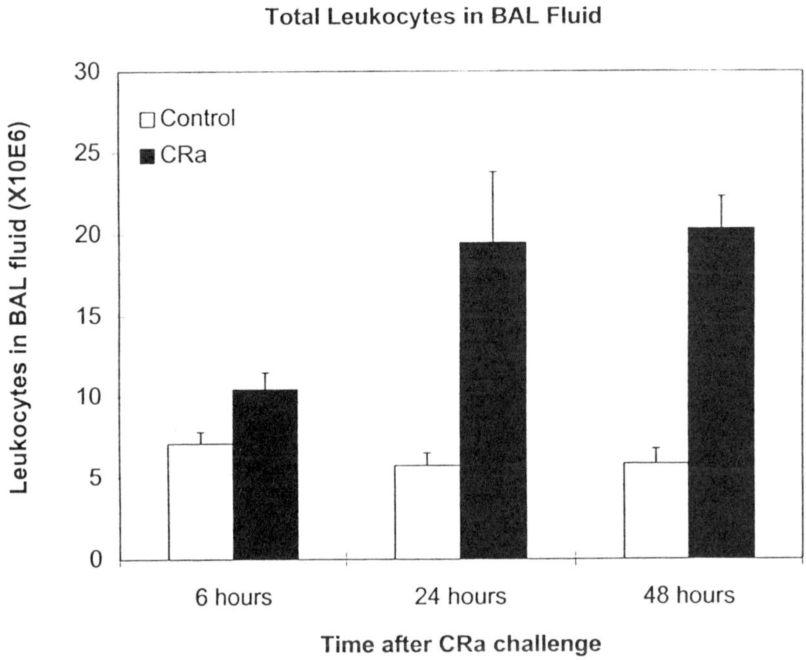

Figure 3. Total inflammatory cells in BALF following CRa provocations. *The numbers are expressed as mean ± SE. Control: Saline immunized (n=6); CRa: CRa immunized (n=6). p<0.01 Control vs. CRa.*

D. INFLAMMATORY CELLULAR RESPONSES

The CRa-sensitive guinea pig model was also examined for its inflammatory cellular responses upon CRa challenges *in vivo*. Inflammatory cells in bronchoalveolar lavage fluid (BALF) at 6, 24 and 48 hours post-CRa challenges were increased in the CRa-immunized guinea pigs as compared with those of control animals (see Fig. 3). Neutrophils and macrophages were major cells at all three time points. When inflammatory cellular responses were compared among three levels of CRa-immunized groups, the overall CRa-provoked cellularity at 24 hour was correlated to the amount of CRa the animals had been exposed to during the immunization period (see

Fig 4). However, the cells that increased in a dose-responsive manner were macrophages and eosinophils. At the beginning we attempted to analyze inflammatory cells in the lungs, fixed cells in the lung parenchyma, separately from the cells in BALF that represent mobile cells. The lung parenchyma was removed after completion of lavage, then chopped and enzyme-digested, and reconstituted, and cells were analyzed. The results were, however, very similar to the pattern of the inflammatory cells in the BALF. Consequently, we stopped performing this separate analysis and stayed with BALF analysis. Eosinophils in the BALF comprised a small

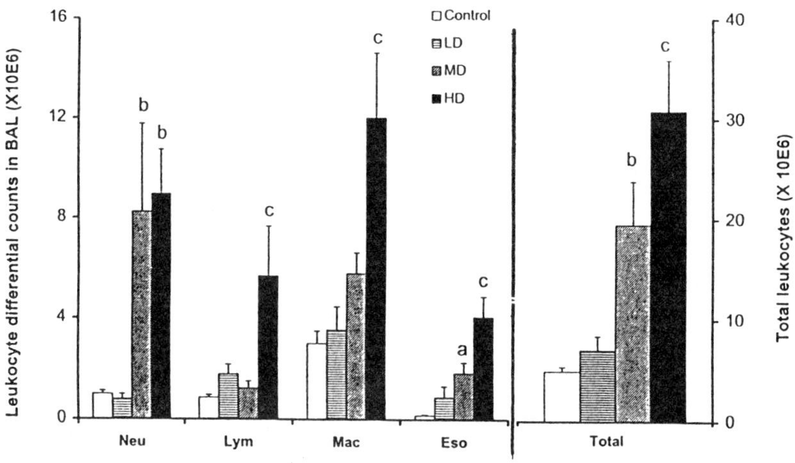

Figure 4. Inflammatory cells and total leukocytes in BALF 24 hours following CRa provocation. *Neutrophils (Neu), monocytes /macrophages (Mac), lymphocytes (Lym), and eosinophils (Eos) in BALF are expressed as Mean ± SE. Control: Saline-immunized (n=11); LD: Low dose (n=6); MD: Medium dose (n=6); HD: High dose CRa-immunized guinea pigs (n=6). a: $p<0.05$ vs. control, b: $p<0.05$ vs. Control and LD, c: $p<0.05$ vs. Control, LD and MD.*

fraction of total cells (3-17%) in the CRa-immunized and challenged animals. However, their increase was related to the CRa doses of immunized animals. The HD guinea pigs immunized with CRa (25 mg/day) showed more eosinophils in the BALF in response to CRa challenge, particularly at 24 hour post-challenge, than did the guinea pigs immunized with LD CRa (0.5 mg/day). However, such leukocytosis or eosinophilia in the CRa-immunized guinea pig does not necessarily correlate with the changes in the airway resistances (SRaw) of the respective groups of guinea pigs in a conscious state. The cellularity in BALF was loosely reflected with increased levels of leukotriene LTC_4/D_4 and LTB_4 in the BALF. We have also attempted to analyze CD_4, CD_8, and a ratio of CD_4/CD_8 to identify a possible marker for obstructive airway response. To date, the results have not been conclusive.

E. PRESENCE OF INFLAMMATORY MEDIATORS

Release of anaphylactic mediators such as histamine, a preformed mast cell mediator and leukotrienes, occurs upon the reaction between allergen and reagins both in humans and animals (IgE, and IgG_1 in the case of guinea pigs). Therefore, presence of the inflammatory mediators such as LTB_4 and LTC_4/D_4 is likely an indication that there are reactions between the CRa and a reagin, IgG_1 antibody, following challenge in the CRa- immunized guinea pigs.

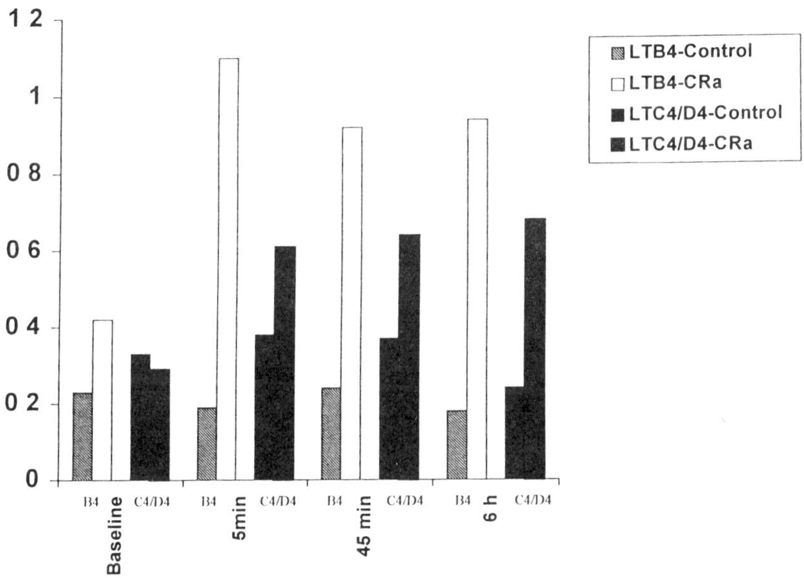

Figure 5. Comparative changes in leukotrienes, LTB4 and LTC4/D4 in arterial blood following CRa-provocation. *Control: Saline (n=8); CRa: CRa-immunized guinea pigs (n=7).*

In the early phase of our efforts to develop the cockroach-asthma model, we documented the presence of elevated leukotrienes, LTB_4 and LTC_4/D_4, in the arterial circulation immediately after CRa inhalation challenge *in vivo* (anesthetized) (Kang et al., 1996). The blood was drawn from the left common carotid artery that was cannulated, one day prior to animals being challenged with CRa. Both LTB_4 and LTC_4/D_4 were elevated at 5, 45 minutes, and at 6 hours while ventilatory function was decreased following CRa challenge in immunized but not in control guinea pigs ($P<0.0001$, Fig. 5). This increase in leukotrienes was further documented in the BALF of the CRa-immunized guinea pig in the next study done on guinea pigs in a conscious state. However, the levels of leukotrienes LTB_4 and LTC_4/D_4 in BALF at 24 hr were not as significant or consistent as those in the carotid arterial blood in the anesthetized state.

In addition, there has been different evidence that the cockroach-sensitive guinea pig model releases anaphylactic mediators upon CRa challenge, specifically histamine and LTD_4. In the airway smooth muscle contractility study, we have documented that the CRa-specific contractions of the airway smooth muscles isolated from CRa-immunized guinea pigs were blocked by an H_1 blocker (pyrilamine) and an LTD_4 receptor antagonist (LY-171883). Figure 6 shows that the CRa-specific contractions of tracheal smooth muscles *in vitro*, expressed in dose-response curves, were markedly diminished when the muscle rings were pre-incubated with Pyrilamine or with LY-171883. The airway muscle strips that had been pre-incubated with a vehicle solution showed unchanged CRa-specific contractions in a dose responsive manner. Pre-incubation with both pyrilamine and LY-171883 almost abolished the cockroach-induced contractility of the tracheal rings. Such observations provide indirect evidence that anaphylactic mediators are responsible for the airway obstructions in response to CRa inhalation documented in the CRa-immunized guinea pigs.

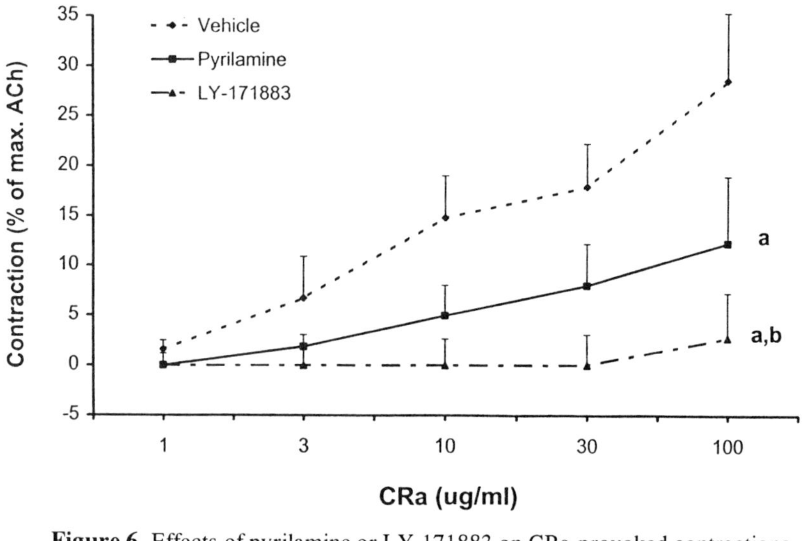

Figure 6. Effects of pyrilamine or LY-171883 on CRa-provoked contractions of tracheal smooth muscle rings. *The tracheal smooth muscle rings were isolated from high dose (HD) CRa-immunized guinea pigs. Data are expressed as a percentage of maximum contraction produced by ACh ($3X10^{-5}$ M) as mean ± SE. a: $p<0.001$ vs. vehicle, b: $p<0.001$ vs. pyrilamime.*

In summary, this guinea pig model created by immunizing animals with aerosolized cockroach allergen alone for a period of 4 weeks shows ample evidence that the sensitization produces development of an anaphylactic antibody, the CRa- specific IgG_1 antibody. A majority of the CRa-immunized guinea pigs undergo asthma-like airway obstruction following inhalation of antigen in the conscious state, as well as under anesthesia. The CRa-provoked airway obstruction occurs *in vivo* at variable time points, thus simulating asthmatic responses in EAR, LAR, or dual patterns. The CRa challenge induces the influx of inflammatory cells in the airways, mostly of neutrophils, macrophages and eosinophils, that lasts for longer than the duration of airway obstruction. The airway smooth muscles of the CRa-immunized guinea pigs respond also with the antigen-specific dose responsive contractions.

V. SIMILARITIES BETWEEN HUMANS AND GUINEA PIGS SENSITIVE TO COCKROACH ALLERGEN

A. PRESENCE OF ANAPHYLACTIC ANTIBODY

The state of the sensitization, as characterized by the presence of a reagin, i.e., anaphylactic antibody (IgE), is most commonly determined by skin test, and sometimes by RAST in humans. Cockroach-sensitive subjects possess marked hypersensitivity as documented by both skin test and RAST (Bernton & Brown, 1967; Mendoza & Snyder, 1970; Kang et al., 1979; Lan et al., 1988; Kang et al., 1989; Kang et al., 1992). The mode of CRa exposure in humans is not clearly documented; however, it is believed that airborne CRa, mostly in the indoor environment, sensitizes humans in a manner similar to other inhalant allergens. In addition to genetic predisposition, a high concentration of the allergen in the inhaled air or a certain threshold amount is likely to induce the sensitization. Likewise, the guinea pigs exposed to airborne CRa chronically became sensitized to CRa, and the immunized state, defined as the presence of anaphylactic antibody, was documented by both skin test and PCA. The anaphylactic antibody in these guinea pigs is IgG_1-like, not IgE antibody. However, it is well known that guinea pigs have two types of anaphylactic antibodies, IgE and IgG_1 (Perini & Mota, 1973; Mapp et al., 1996; Hsiu et al., 1997) and it is apparent that the airborne route of entry of an antigen favors the production of the IgG_1 type antibody (Kang et al., 1995). The presence of IgG_1, in the absence of IgE antibody, is apparently sufficient to cause anaphylactic mediators released upon the CRa challenge *in vivo* as well as *in vitro.* Undem et al., (1986) reported that guinea pig IgG_1 (rather than IgE) was more representative of human IgE as it causes more mediator release and more bronchoconstriction than guinea pig IgE.

B. PRESENCE OF ASTHMATIC RESPONSES UPON CRa INHALATION

As described in section II, humans with cockroach sensitivity respond variably to the inhalation of the cockroach antigen (Kang et al., 1979; Kang et al., 1992). Allergic humans, e.g., subjects have chronic rhinitis and/or sinusitis who show CRa sensitivity by skin test, do not respond with bronchospasm when they do not have asthma (Kang et al., 1979). Some normal subjects also show strongly positive skin test to CRa (Bernton & Brown, 1967; Kang & Sulit, 1978). However, a great majority of atopic subjects with asthma who have shown cockroach sensitivity develop asthmatic responses upon the CRa inhalation challenge (Kang et al., 1979; Kang et al., 1992). The patterns of CRa-provoked asthmatic responses are variable, yet dual responses are a common pattern in cockroach allergic subjects.

In the cockroach-sensitive guinea pigs, a portion of the immunized animals did not respond with airway obstruction to CRa challenge in a conscious state even though sensitivity was confirmed by skin test and PCA (Kang et al., 1997; Zhou et al., 1998). Approximately one half of the inbred and English Shorthair strains and a majority (82%) of the Hartley strain showed mild-to-moderate airway obstructions to a given dose of CRa (Kang et al., 1997). Yet, for the CRa-immunized guinea pigs as a group, the average changes in the ventilatory function in response to CRa challenges showed airway obstruction in the pattern that can be categorized as EAR, LAR, dual, or persistent bronchospasm. Therefore, the airway responses of the CRa-sensitive guinea pigs in a conscious state are similar to the asthmatic responses seen in human cockroach-sensitive asthmatic subjects. Interestingly, however, almost all guinea pigs exposed to CRa aerosols develop a sensitized state by skin test, PCA titers, inflammatory cellularity in BALF, and airway muscle contractions. Moreover, there are signs showing that the sensitized state is dependent on the amount of cockroach antigen guinea pigs are exposed to during the sensitization period (Zhou et al., 1998; Miller et al., 1997). However, the asthmatic responses of these animals, in the conscious state, do not reflect the dose of antigen received during sensitization. In other words, the airway obstruction in the immunized animals in response to a given dose antigen challenge was partially independent of the level of anaphylactic sensitization. Moreover, there was no dose-responsive-effect in the severity of airway obstruction when we compared three dose levels of the CRa immunization in Hartley guinea pigs (see Fig 1). A similar phenomenon has been found in cockroach-sensitive humans. Some cockroach-sensitive subjects, highly allergic to cockroach antigen by skin testing and high RAST titers, do not necessarily have a severe response to CRa challenge (Kang et al., 1992). Clinically, it is also true that not all cockroach-positive (strong) subjects have severe symptoms.

C. PRESENCE OF INFLAMMATORY CELLS IN THE AIRWAYS

The presence of inflammatory cells in the BALF at various time points noted in this guinea pig model cannot be directly compared with the findings of BALF of the asthmatic subjects with cockroach sensitivity, as there are no data on BALF of the cockroach-asthmatic subjects. One should consider the possibility, however, that the asthmatic subjects with cockroach sensitivity might have an inflammatory process of the airways similar to the inflammatory responses found in asthmatic subjects with other inhalants allergies, e.g., dust mite, ragweed, etc. (Smith & Deshazo, 1993; Calhoun et al., 1993; Robinson et al., 1993; Shaver et al., 1995). There have been reports that asthmatic subjects show mildly increased inflammatory cells in the BALF, neutrophils, eosinophils, and lymphocytes, even before antigen provocation. Inhalation of allergen in these asthmatic subjects induces activation of inflammatory cells, while mast cells are degranulated. Leukocytosis with eosinophilia occurs in the BALF of dust mite- allergic asthma subjects after segmental mite challenges, mostly neutrophils and eosinophils, while macrophages were reduced after 24 hour post-challenge (Calhoun et al., 1993; Ying et al., 1995).

Production of leukotrienes in BALF upon CRa challenge noted in this guinea pig model cannot be compared directly with the findings in the human cockroach-sensitive asthmatic subjects, although, this might have been anticipated had one done BALF analysis after CRa challenge in the CRa-sensitive asthmatic subjects.

D. CONTRACTILE RESPONSES OF THE AIRWAY SMOOTH MUSCLES *IN VITRO*

The contractile responses to cholinergic and specific antigen stimulants of the airway muscles of the guinea pig model also cannot be compared directly to the responses of airway smooth muscles *in vitro,* as no study has been done on cockroach-sensitive human airways.

However, it has been reported that the contractile responses of airway smooth muscles isolated from asthmatic subjects with grass sensitivity responded with strong and sustained muscle contractions to grass challenge *in vitro* (Schellenberg & Foster, 1984). There are, however, largely conflicting reports that human asthmatic airways show variable contractile responses to non-specific stimuli such as cholinergic stimuli and B-agonists (Whicker et al., 1988; Bai, 1991; Shaver et al., 1993). Nevertheless, the CRa-specific contractions of the immunized guinea pig airway muscles behaved in a manner similar to the responses of airway smooth muscles represented in other animal asthma models (Compton et al., 1987; Santing et al., 1994; Salari et al., 1994).

VI. CONCLUSION

As presented above we have developed an asthma model in the guinea pig. The model was created by chronic exposure to airborne cockroach allergen, under which condition the guinea pigs develop an IgG_1-like reaginic antibody. The great majority of the sensitized guinea pigs develop "asthma-like" airway obstruction in response to CRa inhalation *in vivo* in a conscious state, a response that is similar to that of humans with cockroach asthma. The model also shows leukocytosis with eosinophilia in the BALF, and its airway muscles show anaphylactic contractions *in vitro*. In many aspects, the model shows similarities with human subjects with cockroach hypersensitivity. Therefore, the cockroach-sensitive guinea pig seems to be an excellent model of asthma for those studies in which humans cannot be used. Studying cockroach-sensitive guinea pigs may increase our fundamental knowledge about the development of human allergic asthma and provide an immunological rationale for the prevention and management of cockroach allergen-induced environmental asthma. As cockroach asthma constitutes a large proportion of high risk asthma, the animal model will provide a tool to explore solutions for this population. Moreover, confirming that an indoor environment and its allergen(s) can induce asthma will foster a better understanding of the development of human asthma, which eventually may lead to better control of the disease.

VII. FUTURE STUDIES PLANNED

Although great strides have been made in asthma research in recent years, the practical difficulties of studying human subjects hinder in-depth investigation of such a common lifelong disease. The availability of this asthma model will be most helpful to elucidate the following: First, contributions and interactions of inflammatory cells such as macrophages, T cells, eosinophils and fixed structural cells such as endothelial and epithelial cells along with fibroblasts, in the development of cockroach sensitization. Studies on the immunological alterations induced in the host while exposed to CRa aerosol will be needed; Second, identification of pivotal differences between the asthmatic state and the allergic state in cockroach allergic guinea pigs; Third, investigation of the mechanisms involved in the neurohumoral control of cockroach allergen induced airway obstruction *in vivo*. The cockroach-sensitive guinea pigs revealed clearly that only a portion of the immunized animals reveal asthma-like responses, thereby, elucidating an underlying mechanism as to why asthma is expressed differently (phenotype) in the CRa-immunized animals; Fourth, the model can also be used to explore various therapeutic modalities for cockroach-allergic asthma. This area could include studies to develop therapeutic agents, such as the use of recombinant allergen, hapten, or antireceptor agents for phase I studies of anti-asthmatic drugs.

Abbreviations used:

CRa	Cockroach allergen
OA	Ovalbumin
EAR	Early asthmatic response
LAR	Late asthmatic response
PCA	Passive cutaneous anaphylaxis
TDI	Toluene diisocyanate
AST	Allergy skin test
SRaw	Specific airway resistance
BAL	Bronchoalveolar lavage
LTB4	Leukotriene, B4
LTC4/D4	Leukotriene C4/D4
EFS	Electric field stimulation
Ach	Acetylcholine

REFERENCES

Anderson P. Antigen-induced bronchial anaphylaxis in actively sensitized guinea pigs. Pattern of response in relation to immunization regimen. *Allergy* 1980; 35: 65-71.

Andersson SE, Zackrisson C, Behrens K, et al. Effect of allergen provocation on inflammatory cell profile and endothelin-like immunoreactivity in guinea pig airways. *Allergy* 1995; 50: 349-358.

Bai TR. Abnormalities in airway smooth muscle in fatal asthma. A comparison between trachea and bronchus. *Am Rev Respir Dis* 1991; 143: 441-3.

Bernton HS and Brown H. Cockroach allergy II: The relation of infestation to sensitization. *South Med J* 1967; 60: 852-5.

Bottomley KMK, Lewis S, Trevor J, et al. A new in vivo method for the measurement of repetitive anaphylactic response in the guinea-pig. *Br J Pharmacol* 1984; 81:195-199.

Bramley AM, Langlands JM, Jones AK, Burgoyne DL, Li Y, Andersen RJ, Salari H. Effects of IZP-94005 (contigma sterol) on antigen-induced bronchial responsiveness in ovalbumum-sensitized guinea pigs. *Brit J Pharmacol* 1995; 114 (8): 1433-8.

Calhoun WJ, Jarjour NN, Gleich GJ, Stevens CA, Busse WW. Increased airway inflammation with segmental versus aerosol antigen challenge. *Am Rev Respir Dis* 1993; 1465-71.

Call RS, Smith TF, Morris E, Chapman MD, Platts-Mills TAE. Risk factors for asthma in inner-city children. *J Pediatr* 1992; 121: 862-6.

Chen G, Zhou D, Kim J, Lee L, Kang BC. Contractile responses to allergen and cholinergic stimulation in bronchus and tracea of cockroach allergen sensitized guinea pigs. Int Arch Allergy Immunol 1998; 116:132-9.

Chen G, Zhou D, Kim J, Wu Z, Lee L, Kang BC. Effects of cockroach antigen (CRa) sensitization on the contractility of airway smooth muscles in asthmatic guinea pigs. *J Allergy Clin Immunol* 1997; 99: S410 (Abstract).

Christiansen SC, Martin SB, Schleicher NC, Koziol JA, Hamilton RG, Suraw BL. Exposure and sensitization to environmental allergen of predominanatly Hispanic children with asthma in San Diego's inner-city. *J Allergy Clin Immunol* 1996; 98: 288-94.

Compton MR, Seale JP, Shaw J. Effects of aerosol antigen challenge *in vivo* on responsiveness of guinea pig tracheal and parenchymal strips *in vitro* to contractile agonists. *Int Arch Allergy Appl Immun* 1987; 84: 40-44.

Evans R, Mullarlly DI, Wilson RW, et al. National trends in the morbidity and mortality of asthma in the United States, prevalence, hospitalization and death from asthma for two decades. 1965:84, *Chest* 1987; 91 (sppl) 655.

Ford RM. Etiology of Asthma. A continuing review (8071 cases seen from 1970-1980). *Ann Allergy* 1983; 50: 47-50.

Gelber LE, Seltzer LH, Bouzoukis JK, Pollart SM, Chapman MD, Platts-Mills TAE. Sensitization and exposure to indoor allergens as risk factors for asthma among patients presenting to hospital. *Am Rev Respir Dis* 1993; 147: 573-578.

Hallsworth MP, Soh CPC, Lane SJ, Arm JP, Lee TH. Selective enhancement of GM-CSF, TNF-a, IL-1ß and IL-8 production of monocytes and macrophages of asthmatic subjects. *Eur Respir J* 1994; 7: 1096-1102.

Hendrix DJ, Davies FJ, D-Sonza MF, Pepys J. An analysis of skin prick test reactions in 656 asthmatic patients. *Thorax* 1975; 30: 2-8.

Hsiu T-R, Lei H-Y, Hsieh AL, Wang TY, Chang H-Y, Chen C-R. Mite induced allergic airway inflammation in guinea pigs. *Int Arch Allergy Immunol* 1997; 112: 295-302.

Humbert M, Durham SR, Kimmitt P, Powell N, Assoufi B, Pfister R, et al. Elevated expression of messenger ribonucleic acid endoding IL-13 in the bronchial mucosa of atopic and non-atopic subjects with asthma. *J Allergy Clin Immunol* 1997; 99: 657-65.

Ishii A, Ito K, Ino Y, Miyamoto T. Experimental asthma in guinea pigs sensitized with mites *(Dermatophagoides farinae)*. *Int Arch Allergy Appl Immunol* 1989; 89: 400-403.

Kang B, Jones J, Johnson J, Kang IJ. Analysis of indoor environment and atopic allergy in urban populations with bronchial asthma. *Ann Allergy*

1989; 62: 30-34.

Kang B, Sulit N: A comparative study on prevalence of skin hypersensitivity to cockroach and house dust antigens. *Ann Allergy* 1978; 41: 333-336.

Kang B, Vellody D, Hamburger H, and Yunginger JW. Cockroach cause of allergic asthma: Its specificity and immunologic profile. *J Allergy Clin Immunol* 1979; 63: 80-86.

Kang BC, Chang JL, Johnson J. Characterization and partial purification of the cockroach antigen in relation to house dust mite (D.F.) antigens. *Ann Allergy* 1989; 63: 207-212.

Kang BC, Wu CW, Johnson J. Characteristics and diagnosis of cockroach sensitive bronchial asthma. *Ann Allergy* 1992; 68: 237-244.

Kang BC, Johnson J, Veres-Thorner C. Atopic profile of inner-city asthma with a comparative analysis on the cockroach-sensitive and ragweed-sensitive subgroups. *J Allergy Clin Immunol* 1993; 92: 1802-811.

Kang BC, Kambara T, Yun DK, Hoppe JF, Lai Y-L. Development of cockroach-allergic guinea pig by simple room air contamination. *Int Arch Allergy Immunol* 1995; 107: 569-575.

Kang BC, Zhou D, Chen G, Kim J. Is there a strain difference in suceptibility of asthmatic inflammation in guinea pigs? *J Allergy Clin Immunol* 1997; 99: S363 (Abstract).

Kang BC, Zhou K, Lai Y-L, Hong CB. Experimental asthma developed by room air contamination with cockroach allergen. *Int Arch Allergy Immunol* 1996; 111: 299-306.

Karol MH. Animal models of occupational asthma. *Eur Respir J* 1994; 7: 555-568.

Koehler PG, Patterson RS, Brewer R Jr. German cockroach infestations in low income apartments *J Econ Entomol* 1987; 80: 446-50.

Lai YL, Zhou K-R. Eglin-c prevents monocrotaline-induced ventilatory dysfunction. *J Appl Physiol* 1997; 82(1): 324-8.

Lai YL: Maximal expiratory flow in the guinea pig. *Lung* 1988; 166: 303-313.

Lan JL, Lee Dt, Win CH, Cheng CP, Yeh CI. Cockroach hypersensitivity: Preliminary study of allergic cockroach asthma in Taiwan. *J Allergy Clin Immunol* 1988; 88: 736-40.

Mapp CE, Lapa e Silva Jr, Luchini RE, Chitano P., et al. Inflammatory events in the blood and airways of guinea pigs immunized with toluene diisocyanate. *Am J Crit Care Med* 1996; 154: 201-8.

Mendoza J, Snyder RD. Cockroach sensitivity in children with bronchial asthma. *Ann Allergy* 1970; 28: 159-63.

Miller B, Zhou D,Gindling H, Profitt R, Sajadi K, Kang B. A comparison between airway resistance and cellular responses in bronchoalveolar lavage fluid (BALF) and in lung interstitium (LUNG) in cockroach allergic guinea pigs. *J All Clin Immunol* 1996; 97:503 (Abstract).

Panlozzi LJ, Coleman JJ, Buist AS. A recent increase in asthma mortality in the Northwestern United States. *Ann Allergy* 1984; 56:392-95.

Patterson R, Grammer LC, Harris KE, Shaughnessy MA. *The role of allergens in airway disease in Asthma and Rhinitis*, ed. Busse and Holgate, Blackwell Science, 1995 pp 895-7.

Perini A, Mota I: The production of IgE and IgG_{1a} antibodies in guinea pigs immunized with antigen and bacterial lipopolysaccharides. *Immunology* 1973; 25: 297-305.

Perrin B, Cartier A, Ghezzo H, et al. Reassessment of the temporal patterns of bronchial obstruction after exposure to occupational sensitizing agents. *J Allergy Clin Immunol* 1991; 87: 630-9.

Platts-Mills TAE, Ward GW Jr, Sporik R, Gelber LE, Chapman MD, Heymann PW. Epidemiology of the relationship between exposure to indoor allergens and asthma. *Int Arch Allergy Appl Immunol* 1991; 94: 339-345.

Pollart SM, Chapman MD, Focco GP, Rose G, Platts-Mills TAE. Epidemiology of acute asthma: IgE antibodies to common inhalant allergens as a risk factor for emergency room visits. *J Allergy Clin Immunol* 1989; 83: 875-82.

Robinson D, Hamid J, Bently A, Ying S, Kay AB, Durham SR. Activation of CD_4+ T cells, increased TH_2-type cytokine in RNA expression and eosinophil recruitment in bronchoalveolar lavage after antigen inhalation challenge with atopic asthma. *J Allergy Clin Immunol* 1993; 92: 313-24.

Rosenstreich DL, Eggleston P, Kattan M, et al. The role of cockroach allergy and exposure to cockroach allergen in causing morbidity among inner-city children with asthma. *N Engl J Med* 1997; 336: 1356-63.

Roth LM, Wilis ER. The medical and verterinary importance of cockroaches. *Smithsonian Miscellaneous Collection* 1957; 134: 1-149.

Salari H, Howard S, Chan H, Dryden P, Chan-Yeung M. Involvement of immunologic mechanisms in a guinea pig model of western red cedar asthma. *J Allergy Clin Immunol* 1994; 93: 887-84.

Santing RE, Hoekstra Y, Pasman Y, Zaagsma J, Meurs H. The importance of eosinophil activation for the development of allergen-induced bronchial hyperreactivity in conscious unrestrained guinea pigs. *Clin Exp Allergy* 1994; 24: 1157-1163.

Santing RE, Olymulder CG, Zaagsma J, Meurs H. Relationships among allergen-induced early and late phase airway obstructions, bronchial hyperreactivity, and inflammation in conscious, unstrained guinea pigs. *J Allergy Clin Immunol* 1994; 93: 1021-30.

Santing RE, Schraa EO, Vos BG, Gores RJ, Olymulder CG, Meurs H, Saagsma J. Dissociation between bronchial hyperreativity *in vivo* and reduced ß-adreno-receptor sensitivity *in vitro* in allergen-challenged guinea pigs. *Euro J Pharmacol* 1994; 257: 145-52.

Sarpong SB, Hamilton RG, Eggleston PA, Adkinson ND jr. Socioeconomic status and race as risk factors for cockroach allergen exposure and sensitization in children with asthma. *J Allergy Clin Immunol* 1996; 97: 1393-1401.

Sarpong SB, Wood RA, Karrison T, Eggelston PA. Cockroach allergen (Bla g 1) in school dust. *J Allergy Clin Immunol* 1997; 99: 486-92.

Sastre J, Ibanez MD, Lombardero M, Laso MT, Lehrer S. Allergy to cockroaches in patients with asthma and rhinitis in urban area (Madrid). *Allergy* 1996; 51: 582-6.

Schellenberg RR, Foster A. *In vitro* responses of human asthmatic airway and pulmonary vascular smooth muscle. *Int Arch Allergy Apply Immunol* 1984; 75: 237-41.

Schellenberg RR, Evidence of smooth muscle dysfunction in asthma. In: *Asthma and Rhinitis* edited by WW Busse and SJ Holgate, Blackwell Science, 1995, pp 1150-1156.

Seth KK, Lemanske Jr., RF, The early and late asthmatic response to allergen, *IBID*, pp 946-960.

Shaver JR, O'Connor JJ, Pollice M, et al. Pulmonary inflammation after segmental ragweed challenge in allergic asthmatic and non-asthmatic subjects. *Am J Respir Crit Care Med* 1995; 152: 1189-97.

Smith DL, Deshazo RD, Bronchoalveolar Lavage an Update and Perspective. *Am Rev Respir Dis* 1993; 148: 523-32.

Southrada M, Southrada JF. Reassessment of electrophysiological and contractile characteristics of sensitized airway smooth muscle. *Respir Physiol* 1981; 46: 17-27.

Sperber K, Kendler D, Yu LM, Nayak H, Pizzimenti A. Prevalence of atopy in an inner-city asthmatic population. *Mt. Sinai J Med* 1993; 60: 227-31.

Sporik R, Holgate ST, Platts-Mills TAE, Cogswell JJ. Exposure to House Dust-Mite Allergen (DER p I) and the Development of Asthma in Childhood. *N Engl J Med* 1990; 323: 502-507.

Steinberg DR, Bernstein DI, Gallagher JS, Arlian L, Bernstein IL. Cockroach sensitization in laboratory workers. *J Allergy Clin Immunol* 1987; 80: 586-590.

Sugiyama H, Okada C, Bewtra AK, Hopp RJ, Townley RG. The effect of formoterol on the late asthmatic phenomena in guinea pigs. *J Allergy Clin Immunol* 1992; 89:858-866.

Undem BJ, Buckner CK, Harley P, Graziano FM. Smooth muscle contraction and release of histamine and slow-reacting substance of anaphylaxis in pulmonary tissues isolated from guinea pigs passively sensitized with IgG_1 or IgE antibodies. *Am Rev Respir Dis.* 1985; 131: 260-266.

Weiss KB. Seasonal trends in US asthma hospitalizations and mortality. *JAMA* 1990; 263: 2323-28.

Whicker SD, Armour CL, Black JL. Responsiveness of bronchial smooth muscle from asthmatic patients to relaxant and contractile agonists. *Pulmonary Pharmacology* 1988; 1: 25-31.

Ying S, Durham SR, Corrigan CJ, Harnid P, Kay AB. Phenotype of cells expressing in RNA for TH_2-type (interleukin 4 and interleukin 5) and TH_1-type (interleukin 2 and interfern 2) cytokines in bronchoalveolar lavage and bronchial biopsies from atopic asthmatic and normal control subjects. *Am J Respir Cell Mol Biol* 1995; 12: 477-87.

Zhou D, Chen G, Kim J, Kang B. A dose relationship to cockroach allergen (CRa) sensitization and asthma in Hartley Guinea Pigs. *J Allergy Clin Immunol* 1997; 99:S500 (abstract).

Zhou D, Chen G, Kim J-T, Lee L-Y, Kang B. A dose-response relationship between exposure to cockroach allergens and induction of sensitization in an experimental asthma in Hartley guinea pigs. *J Allergy Clin Immunol* 1998; 101: 653-9.

Chapter 8

THE EFFECTS OF ENVIRONMENTAL TOBACCO SMOKE ON ASTHMA: STUDIES WITH A DYNAMIC EXPOSURE CHAMBER

Samuel B. Lehrer, Ph.D., Roy J. Rando, Sc.D., Manuel Lopez, M.D.

I. INTRODUCTION

Cigarette smoking has declined in the United States from 42.4% of the population in 1965 to 25.5% in 1994. Approximately 50 million Americans smoke, increasing the likelihood that a number of non-smoking American adults and children may be passively exposed to Environmental Tobacco Smoke (ETS) (Rando et al, 1997). Although smoking of tobacco products is a significant health risk for smokers, the health effects of ETS on non-smokers exposed passively have only recently become of interest (Rando et al, 1997; Shepard et al, 1979; Wiedemann et al, 1986; Dahms et al, 1981; Chilmonczyk et al, 1993; Brownson et al, 1992). Increased concern by the American public with regard to potential negative effects about passive smoke exposure has resulted in the banning of smoking in public places (US Dept. Health & Human Services, 1991; Centers for Disease Control, 1992).

Tobacco smoking has long been recognized as a significant health risk for smokers. Smoke generated by burning cigarettes can be separated into mainstream smoke which the smoker inhales and sidestream smoke which emanates from the burning tip of the cigarette. ETS consists of exhaled mainstream and side stream smoke (Centers for Disease Control, 1992). A variety of health effects have been demonstrated in patients exposed to ETS. It can induce ocular, upper and lower respiratory symptoms in exposed nonsmokers, and increases in the number of allergic individuals reporting irritating symptoms (Menon et al, 1992; Danuser et al, 1993). In children, exposure to smoke has been associated with increased acute lower respiratory infections, chronic cough, persistent otitis media (Kraemer et al, 1983) and decrease in lung function, increased airway hyperreactivity, and increased number of visits to the emergency room with asthma exacerbations (Murray & Morrison, 1986, 1988; Evans et al, 1987).

Studies examining the effects of short term exposure to tobacco smoke on respiratory diseases such as asthma have yielded conflicting results. Two studies (Shepard et al, 1979; Wiedemann et al, 1986) did not show a decline in pulmonary function of test subjects following ETS exposure; whereas others, (Dahms et al, 1981; Stankus et al, 1988; Menon et al, 1988) have demonstrated a significant effect on lung function in some asthmatics. These previous studies were performed using a static smoke chamber. However, static chambers do not always reflect the typical smoke atmospheres to which the public is exposed (Lehrer et al, 1997).

1-56670-373-5/99/$0.00+$.50

This report will describe a dynamic test challenge chamber developed to conduct ETS exposure studies under more realistic and more precise conditions. The effects of ETS on pulmonary and non-pulmonary responses, reproducibility of challenge and sensitivity of exposure, as well as, identification of active ETS components and potential mechanisms involved in ETS-induced bronchospastic responses are addressed.

II. DEVELOPMENT/EVALUATION OF A DYNAMIC SMOKE EXPOSURE CHAMBER

ETS is an exceedingly complex mixture of chemical constituents that are emitted into the atmosphere in both particulate and vapor phases (Table 1). Many of the components of ETS are unstable and degrade over time through chemical reactions and physical removal processes, including oxidation, hydrolysis, surface adsorption, and particle sedimentation. Thus, as ETS ages, its qualitative and quantitative composition is constantly changing. As a result, careful attention must be paid to the techniques used for producing test atmospheres in order to accurately reproduce realistic conditions of exposure.

The dynamic challenge chamber, when operated at a ventilation rate representative of typical indoor environments, is the most satisfactory approach to production of test atmospheres of ETS. In terms of the chemical composition of the ETS atmosphere, the dynamic challenge chamber provides a "snapshot" in time representative of a precise age of ETS which is indefinitely stable as long as the chamber conditions are maintained. In contrast, in the static challenge chamber, the atmospheric composition is continuously changing over time. If cigarettes are smoked at regular intervals in the static chamber, certain constituents, such as carbon monoxide, will build up in concentration over time, whereas others, such as particulate levels, may remain relatively constant or decrease.

In this study, a dynamic challenge chamber was constructed by modifying an existing office room. The dynamic challenge chamber allowed very precise control of test atmospheres for as long as necessary by varying the number of cigarettes smoked and adjusting the air flow to assure accurate levels of ETS (Rando et al, 1992). The chamber's design is diagrammed in Figure 1. The chamber enables exposure of study subjects to passive cigarette smoke under "real-life conditions". Development and testing of this chamber have been described previously (Rando et al, 1992). The chamber consists of an antechamber that houses smoke generation equipment and a central station for monitoring devices and spirometer. Smoke is carried into the challenge chamber (2.57 x 2.49 x 2.64 meters, volume 16.9 cubic meters) through an air diffuser grill in the common wall between the chamber and the antechamber. The flow rate in the chamber is controlled by a combination of adjustable louvers in the exhaust duct and the two-speed

Table 1.
Emission Factors for Mainstream and Sidestream Tobacco Smoke

	mainstream (μg per cigarette)	sidestream (μg per cigarette)
carbon dioxide	10,000-80,000	81,000-640,000
carbon monoxide	500-26,000	1200-65000
oxides of nitrogen	16-600	80-3500
ammonia	10-130	400-9500
hydrogen cyanide	280-550	48-203
formaldehyde	20-90	1000-4600
acrolein	10-140	100-1700
N-nitrosodimethylamine	0.004-0.18	0.04-149
nicotine	60-2300	160-7600
total particulate	100-40,000	130-76,000
phenol	20-150	52-390
catechol	40-280	28-196
naphthalene	2.8	45
benzo(a)pyrene	0.008-0.04	0.02-0.14
aniline	0.10-1.20	3-36
2-naphthylamine	0.004-0.027	0.02-1.1
4-aminobiphenyl	0.002-0.005	0.06-0.16
N-nitrosonornicotine	0.2-3.7	0.02-18

Source: Guerin, et al., Measuring environmental emissions from tobacco combustion: Sidestream cigarette smoke literature review. Atmos Environ 1987;21:291-297

controller in the blower. The chamber could be ventilated at a rate of 1.7 to 10.8 m^3/min (0.1 to 0.7 air changes per minute, Rando et al, 1992).

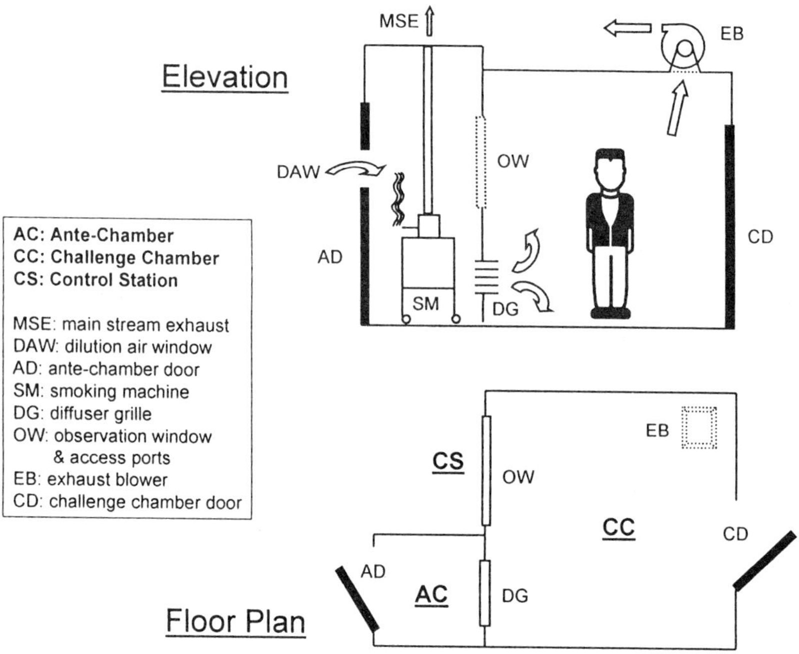

Figure 1: Schematic diagram of ETS Challenge Chamber. The chamber consists of an ante-chamber (AC) where smoke is generated, challenge chamber (CC) where the subject is placed, and a control station (CS) where the equipment for lung function and environmental measurements is contained.

In order to characterize the test atmospheres of SS-ETS produced in the chamber, various markers of tobacco smoke were measured. Evaluations were conducted over a range of ETS levels produced by varying the chamber ventilation rate while one or two University of Kentucky 1R4F research cigarettes were continuously smoked with a Borgwaldt automatic smoking machine. The cigarettes were smoked at a rate of one 75-mL puff per minute and each cigarette lasted 8 minutes after which it was extinguished and immediately replaced with a fresh, burning cigarette. For each set of chamber operating conditions, a "smoke index (SI)", equivalent to an overall ETS concentration, was calculated as:

$$SI = \frac{G}{Q}$$

Table 2.
Characterization of side-stream environmental tobacco smoke atmospheres produced in the dynamic challenge chamber

Chamber Flow (m^3/min)	# of cigarettes smoking	Smoke Index (Cig-min/m^3)	UVPM ($\mu g/m^3$)	Nicotine ($\mu g/m^3$)	CO (ppm)	NO_2 (ppb)	NH_3 (ppb)	Formaldehyde (ppb)	Acetaldehyde (ppb)	Acrolein (ppb)
1.7	2	1.18	1553 ± 53	379 ± 13	12	164 ± 12	930 ± 45	159 ± 17	162 ± 11	33.0 ± 2.6
1.7	1	0.59	621 ± 36	216 ± 7	8	108 ± 13	525 ± 41	77 ± 1	56 ± 3	8.1 ± 0.2
7.4	2	0.27	337 ± 4	148 ± 7	6	59 ± 16	383 ± 26	35 ± 0.4	29 ± 0.5	4.6 ± 0.6
7.8	1	0.13	121 ± 24	63 ± 6	1	ND	80 ± 17	19 ± 2	18 ± 2	3.6 ± 0.3
10.8	1	0.09	86 ± 4	45 ± 3	ND	ND	110 ± 15	9 ± 1	7 ± 0.2	2.5 ± 0.6
*Generation Factor (μg/cigarette)			2358	341	13,112	283	579	164	229	58

ND: not determined

* Generation factor equivalent to slope of zero-intercept linear regression of pollutant concentration v. smoke index.

Lehrer et al, Indoor Air, 1997;7(4):269-277.

Where G, the "generation rate", is the number of cigarettes being smoked, and Q is the chamber dilution flow rate.

Table 2 presents the results of characterization of the chamber for various markers of ETS (ultraviolet absorbing particulate matter [UVPM], nicotine, carbon monoxide) as well as certain respiratory irritants. The measured levels of these pollutants in the test atmospheres are highly linearly correlated with the smoke index which reflects on the reproducibility and predictability of the chamber. The generation factors for the various pollutants calculated from the characterization data are consistent with data reported in the literature for emissions from cigarettes (Table 1).

III. EFFECTS OF ETS ON PULMONARY AND NON-PULMONARY RESPONSES

A. DYNAMIC CHALLENGE CHAMBER EXPOSURE

The exposure of study subjects in a dynamic challenge chamber has recently been described (Lehrer et al, 1997). Briefly, all subjects underwent a standard inhalation challenge to ETS at a smoke index of 1.18 cigarettes - minute/meter^3 which corresponds to UVPM of 1553 $\pm$ 53 $\mu g/m^3$, nicotine (NIC) level of 379 $\pm$ 13 $\mu g/m^3$ and a carbon monoxide level of approximately 12 ppm (Lehrer et al, 1997). Any subject demonstrating a decline in their lung function i.e. decrease in forced expiratory volume (FEV_1) $\geq$ 20%, was designated as a reactor and allowed to exit the test chamber at that time. Subjects were monitored for up to 24 hours (except when asleep) following commencement of the challenge with hourly peak flow measurements using a Miniwright peak flow meter. Subjects completed a symptoms response sheet upon completion of the challenge in which they recorded conjunctiva, nasal, pharyngeal or lower respiratory tract symptoms and overall acceptability or unacceptability of the exposure.

A total of 130 asthmatics were exposed to ETS. They ranged from 18 to 60 years old (mean age 34), 50% were Caucasian, 50% were black, 71% were female, and most (86%) were atopics. Of the 130 asthmatics challenged, 26 (20%) demonstrated a significant decrease in FEV_1 (Lehrer et al, 1997). Most of the subjects reacted after 90-120 minutes (mean 120 min) of exposure. All reactors to ETS challenge were subjected to a sham challenge control exposure without the presence of cigarette smoke (for the same duration of the smoke challenge). This control challenge was performed on all reactors to determine the specificity of reactivity. Of 26 ETS reactors, 6 had positive sham challenge responses. Examples of four reactor responses (2 with and 2 without positive sham reactions) are shown in Figure 2.

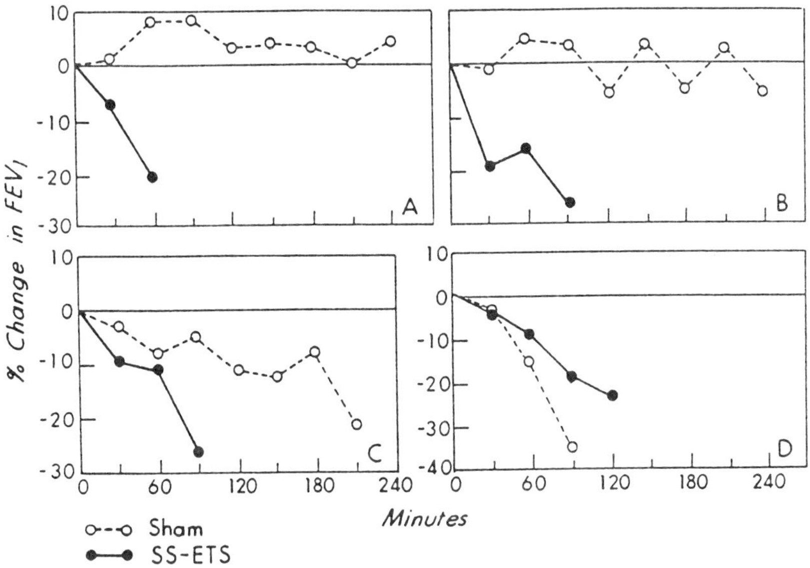

Figure 2: Lung function changes as measured in FEV_1 of asthmatics exposed to ETS or sham challenge.
All subjects had a positive $\geq$ 20% drop in FEV_1 to ETS (SI 1.18, UVPM 1553 $\pm$ 53 $\mu g/m^3$). Subjects A and B had a negative sham challenge; subjects C and D also had positive sham challenges.

Most asthmatics exposed to ETS reported moderate to severe non-pulmonary symptoms such as nasal, conjunctival, or pharyngeal irritant symptoms following smoke exposure. However, no relationship was detected between the presence or severity of non-pulmonary symptoms and the subject's objective responses or own perceived smoke "allergy" (Table 3). Perceived smoke allergy was significantly different when asthmatics were compared to non-asthmatics ($p=0.001$) or ETS reactor asthmatics were compared to ETS non-reactor asthmatics ($p=0.037$, 1-tailed t-test).

B. DURATION OF ETS REACTIVITY

The persistence of ETS reactivity was also determined. Although these studies were performed in a static inhalation challenge chamber, they serve to illustrate that ETS reactivities apparently are long lasting. Six reactors who were demonstrated to have a significant ($\geq$20%) decline from baseline FEV_1 were exposed 24 months later to ETS (Menon et al, 1991). Five of the 6 reactors still demonstrated substantial reactivity to ETS; none of the 9 previously demonstrated non-reactors had any reactivity to ETS upon exposure 24 months later.

Table 3.
Relationship of perceived smoke allergy to non-pulmonary symptoms

Perceived smoke allergy	Asthmatics[a]			Controls (non-asthmatics) n=19
	ETS - n=79	ETS + Sham - n=14	ETS (-)+ Sham + n=6	
Yes	59 (75%)[b]	13 (93%)	6 (100%)	7 (37%)
No	20 (25%)	1 (7%)	0 (0%)	12 (63%)

Lehrer et al, Indoor Air, 1997;7(4):269-277.

(a) Asthmatics are classified non-responders (ETS-), specific responders (ETS+ and, sham-) or non-specific responders (ETS+ and Sham+).

(b) Figures are the number of individuals reporting non-pulmonary symptoms, i.e. nasal, eye or pharyngeal irritation.

Table 4. Responses of smoke reactors to diminishing levels of ETS

Subject	1.18[1]		0.59		0.27		0.128	
	FEV_1 drop		FEV_1 drop		FEV_1 drop		FEV_1 drop	
	%	Minutes	%	Minutes	%	Minutes	%	Minutes
1	-24	180[2]	Negative					
2	-30	90	Negative					
3	-20	240	Negative					
4	-21	30	-24	180	Negative			
5	-20	150	-41	150	-41	90	Negative	
6	-37	120	-22	30	-33	210	-38	180
7	-35	120	-24	180	-20	90	-22	90

[1] SS-ETS, smoke index=number of cigarettes continuously smoked ÷ dilution flow rate.

[2] % FEV_1 drop/minutes.

Lehrer et al, Indoor Air, 1997;7(4):269-277.

C. LEVELS OF ETS THAT INDUCE REACTIVITY

The levels of ETS to which individuals react is of growing medical legal interest. Our initial studies tested relatively high (SI=1.18) levels of ETS typical of bars and casinos where substantial smoking occurs. Since such levels, particularly in this current period of reduced smoke exposure, reflect high levels of ETS exposure, it was important to assess the reactivity of subjects to lower ETS levels. Therefore, 7 reactors identified in our dynamic chamber challenge studies were exposed to diminished levels of ETS according to the established procedure (Lehrer et al, 1997). Results (summarized in Table 4 and Figure 3) indicate that 4 of the 7 individuals responded to diminished ETS levels.

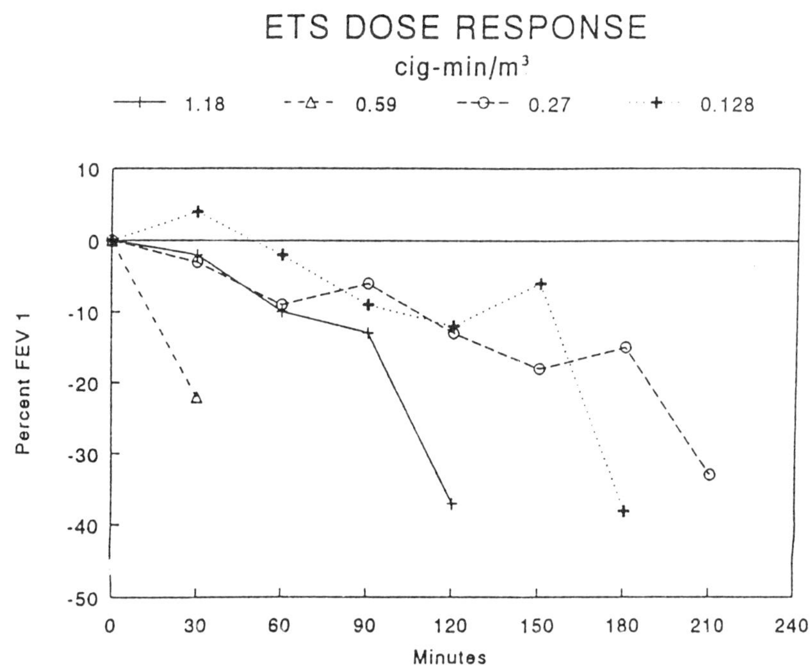

Figure 3: ETS Dose Response.
The responses (as measured in drop of percent of FEV 1) of an asthmatic to reduced levels of ETS are shown. This subject reacted to the lowest ETS level demonstrating substantial reactivity. There was no correlation of the levels of ETS exposure with the length of time to respond.

Two of the 4 responders reacted to ETS levels that were the lowest possible we were able to generate. The levels of ETS used in this study ranged from a smoke index of 1.18 to a low of 0.13 cig-min/m^3 (Table 2). The upper limit of 1.18 cig-min/m^3 represents a high exposure level which would be encountered in the real world only in the smokiest environments. In contrast, the UVPM and nicotine concentrations measured in the chamber at the low smoke index of 0.13 cig-min/m^3 are similar to the average levels of these ETS markers reported in surveys or peak levels measured in the homes of smokers during smoking cycles (Rando, 1997).

These results indicate that although the 15.5% of asthmatics who react to ETS is a lower proportion than expected, some of these individuals are exquisitely sensitive to ETS exposure with regard to triggering their asthma. Responders to these low levels of ETS reflect a small (<8%) portion of the total population of asthmatics who complain of symptoms upon ETS exposure.

IV. MECHANISMS OF ETS-INDUCED BRONCHIAL CONSTRICTIVE RESPONSES & IDENTIFICATION OF ACTIVE ETS COMPONENTS

A. MECHANISMS

In the past, some physicians proposed that tobacco smoke allergy exists in many allergy patients as demonstrated by positive skin test to tobacco leaf extracts and it has also been suggested that approximately 20% of the American population is allergic to tobacco smoke (Harkavy, 1968; Taylor, 1974; Zussman, 1974; Becker et al, 1976). Since large numbers of individuals were reported to have positive skin test reactivity to tobacco leaf extracts and some of these subjects believe they are allergic to tobacco smoke, the possibility that IgE antibody reactivity to tobacco allergens causes tobacco smoke reactivity was investigated. Subjects undergoing tobacco smoke challenges were skin prick tested with a number of commercially and laboratory prepared tobacco leaf extracts, as well as extracts of ETS in phosphate buffered saline (Menon et al, 1987; Menon et al, 1991; Musmand et al, 1992). Although a significant number of subjects had positive skin test reactions to at least one of the tobacco leaf extracts tested, and lesser numbers reacted to tobacco smoke extracts, no correlation (Fisher's Exact Test) between skin test reactivity to tobacco leaf or smoke extracts with any objective pulmonary response to ETS was demonstrated (Table 5). This strongly suggests that reactivity to ETS is not caused by an IgE-mediated mechanism.

To investigate the potential pathophysiology further, nasal secretions induced by acute exposure to ETS were studied (Musmand, White, Lopez, et al, 1993). Twelve atopic asthmatics were exposed to ETS as described previously and nasal secretions were collected utilizing the nasal lavage procedure developed at the National Institutes of Health (Raphael et al, 1988; Raphael et al, 1989). Total protein, albumin, and lactoferrin were quantitated. Four of 12 study subjects manifested $\geq$20% drop in FEV_1

Table 5.
Skin test results to tobacco and smoke extracts[a]

	POSITIVE CHALLENGE SUBJECTS	NEGATIVE CHALLENGE SUBJECTS
ATOPIC	7/8	43/50
COMMERCIAL TOBACCO MIX EXTRACT	2/8	11/50
TOBACCO LEAF EXTRACT	1/8	7/50
SMOKE EXTRACT	2/8	2/50

[a] All subjects were tested by skin prick test method and wheal diameters determined 15 minutes later (largest diameter and one perpendicular to it). Mean wheal diameters $\geq$ 3 mm are considered positive.

Table 6.
Environmental tobacco smoke challenge studies of asthmatics

CGRP release (pg/ml)			
	Baseline	Maximum	
Nonreactors	16.0 ± 3.8	48.8 ± 4.7	$p<0.001$
Reactors	15.2 ± 11.2	58.5 ± 3.1	$p=0.009$
	$p=0.929$	$p=0.227$	
VIP release (pg/ml)			
	Baseline	Maximum	
Nonreactors	16.9 ± 3.8	33.9 ± 3.9	$p=0.005$
Reactors	12.1 ± 4.3	14.0 ± 3.6	$p=0.738$
	$p=0.449$	$p=0.013$	

during the exposures. All 12 reported significant increase in subjective ocular, nasal, lower respiratory symptoms and overall symptoms. Levels of total protein and albumin in the nasal lavage fluid did not change significantly during the challenges, indicating that there was no significant increase in vascular permeability from acute ETS exposure. Similarly, levels of lactoferrin did not vary significantly during the challenge, suggesting that acute ETS exposure did not include significant amounts of glandular secretion. Levels of the proteins assayed did not differ between the group of subjects with or without a significant drop in FEV_1. Consequently, respiratory tract responses to acute ETS exposure are not consistent with an allergic reaction or a cholinergic mediated response.

The role of neuropeptides in respiratory tract response to ETS was also studied, by testing nasal lavage fluids obtained as described above for neuropeptides. Neuropeptides are important in responses to sensory stimulation by having potent effects on vascular and bronchomotor tone; airway secretions and inflammatory reaction (Barnes et al, 1991a,b). Levels of calcitonin gene-related peptide (CGRP) and vasoactive intestinal peptide (VIP) were measured in the nasal secretions of the 12 asthmatics acutely exposed to passive smoke as described (Musmand, White, Mosimann, et al, 1993). During the challenges, total suspended particulate matter and nicotine levels averaged 1502 $\pm$ 300 $\mu g/m^3$ and 399 $\pm$ 114 $\mu g/m^3$, respectively. Four of the 12 subjects manifested a significant decline in FEV_1. Levels of CGRP, a potent bronchoconstrictor postulated to be the predominant mediator of vasodilation in response to sensory neural stimulation in the nasal and bronchial mucosa, increased from baseline in the nasal secretions of the challenge subjects during most of the time periods sampled. Levels did not differ between those with a significant fall in FEV_1 and those without such a decline. Levels of VIP, which has been found to cause vasodilation and have potent bronchodilitory effects on airway tone, although elevated significantly in asthmatics with no significant drop in FEV_1, did not change significantly from baseline in subjects with a significant decline in FEV_1 (Table 6), suggesting that asthmatics' respiratory tract response to acute ETS exposure can, in part, be explained by perturbations in neuropeptide secretion.

B. ACTIVE ETS COMPONENTS

What in ETS induces bronchial restrictive responses? Although neither the mechanism(s) nor the active agent(s) have been identified, there is some information as to the nature of ETS components that cause bronchial constriction. Initial studies separated particulates from the gaseous elements in ETS to determine which components may be important in bronchoconstriction. Several filters were tested for particulate removal and the resulting gaseous phase analyzed.

The Astrocell HCX filter (American Filter Co., Louisville, KY) consistently exhibited filtration efficiencies for smoke particulate (UVPM) in the range of 92 to 98%, while passing 65 to 100% of measured gas phase components including nicotine, formaldehyde, acetaldehyde, acrolein and ammonia. Thus, the Astrocell HCX filter was installed between the antechamber and challenge chamber and used to produce filtered smoke atmospheres for use in challenge testing of gas phase ETS.

Using the challenge chamber with the Astrocell HCX filter, 7 previously demonstrated ETS reactors and 6 non-reactors were exposed to filtered ETS (gaseous phase, particulates removed) until either a significant drop in FEV_1 occurred or a 4-hour exposure elapsed. For these 13 filtered ETS challenges, UVPM, nicotine and carbon monoxide were measured as markers of exposure. Table 7 indicates that the level of ETS particulate is reduced by about 91% in comparison to whole, unfiltered ETS atmosphere at the same smoke index. Nicotine is reduced about 65%, whereas carbon monoxide, which is completely in the gas phase and is essentially inert, remains unchanged.

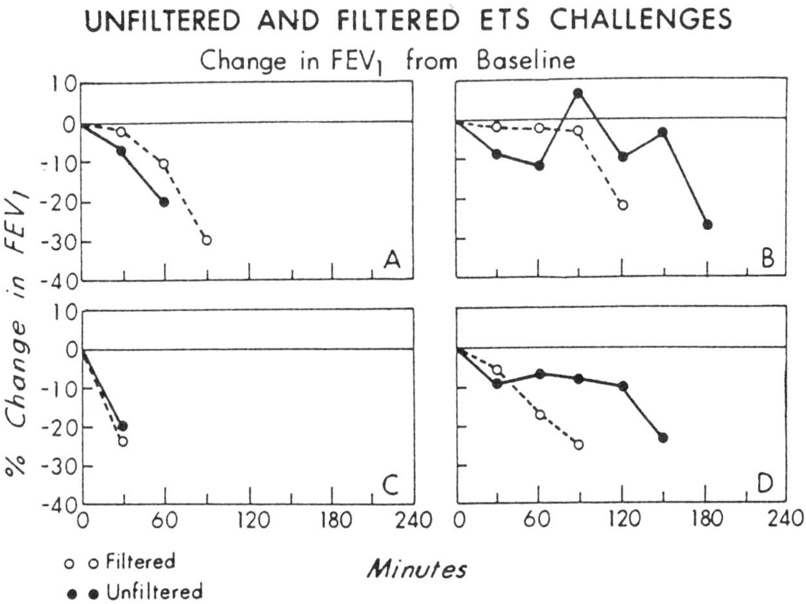

Figure 4: Comparison of reactivity of asthmatics (as measured by change in FEV_1) to filtered (gaseous phase) or unfiltered ETS. Four of the 5 subjects whose response to filtered ETS was similar to their responses to unfiltered ETS are shown here. Predicted unfiltered ETS levels for UVPM were 1484 $\mu g/m^3$ (Table 7). Filtered ETS UVPM levels were 135 $\pm$ 103 $\mu g/m^3$. Other components are listed in Table 7.

Five of the 7 reactors had positive responses to the gaseous phase exposure whereas only 1/5 non-reactors had a drop similar in magnitude

(Figure 4). Four of these 5 reactors had a positive challenge response at 30-60 minutes following a gaseous phase exposure. These results demonstrate that most ETS reactors respond to components present in the gas phase and strongly suggests that the non-particulate, i.e., gaseous components, may be the primary etiologic agents in bronchial constrictive responses in asthmatics. Certainly, these components warrant further study and may help guide attempts to produce cigarettes with reduced adverse health effects by reduction or abolishing these components.

Table 7.
Markers of side-stream ETS in filtered-smoke (gaseous Phase) challenge atmospheres

	Predicted Level in Whole Smoke (1.18 cig-min/m^3)	Measured Level in 13 Filtered-Smoke Challenges
UVPM ($\mu g/m^3$)	1484	135 $\pm$ 103
Nicotine ($\mu g/m^3$)	402	142 $\pm$ 18
CO (ppm)	13.5	13.6 $\pm$ 2.5

V. SUMMARY AND CONCLUSIONS

Investigations in a dynamic exposure chamber (with proper controls) indicate that only 15% of asthmatics who complain of ETS reactivity demonstrate significant objective changes in lung function upon exposure to ETS. Why most asthmatics complain of ETS reactivity yet only a small percentage have documented bronchoconstriction is not clear. Perhaps this could be due to other symptoms that most asthmatics have such as conjunctivitis, rhinitis, or cough that they may attribute as worsening of their asthma. Another possibility is that even though many have less than a 20% drop in lung function, such a drop in individuals with compromised airways could be clinically significant.

Clearly, when undertaking such studies, it is essential to have exposures that mirror real life situations. This is the advantage of our current studies with the dynamic challenge chamber, testing reduced levels of ETS that are reflective of real life exposure levels. Such studies indicate that the percentage of individuals who respond to low ETS levels is much smaller (8%) than that (15%) responding to higher level exposures. This also demonstrates that some individuals can be exquisitely sensitive to ETS bronchoconstrictive reactivity.

There are modifications in exposure to mimic other real life situations. For example, Peden's group (Peden et al, 1997) have investigated the bronchial response to ozone exposure in asthmatic individuals. They have demonstrated significant bronchoconstrictive responses when exercise occurs during the exposure regimen. Certainly, exercise appears to be an important component of some forms of asthma. However, although this may be an important approach for future investigations of ETS effect on bronchoconstriction in asthma, it is unusual for significant exercise to be performed in the presence of elevated ETS levels.

Finally, the mechanism of ETS exposure and the identity of etiologic agents present in ETS have not yet been resolved. Clearly, ETS reactivity is not a classical allergic reaction. The dynamics of the response to ETS do not favor an immediate or late phase IgE mediated reaction and there is no association of skin test reactivity to tobacco leaf or smoke extracts with bronchoconstrictive responses to ETS. More likely, positive skin test reactivity to tobacco leaf extracts reflects cross-reactivity with other plant allergens to which the subject is sensitized. Finally, the nature of the ETS components that induce bronchoconstriction is not resolved although they do not appear to be particulate substances. This suggests that gaseous components are responsible for the effects seen. The fact that there appears to be imbalance in mediator release in nasal lavage fluids in asthmatics responding to ETS suggests that neuropeptides may play a role in symptoms induced by exposure to ETS. These observations suggest methods by which the respiratory effects of ETS could be reduced (i.e. lower gaseous components) and treatment for symptomatic asthmatics who react to ETS (i.e. neuropeptide blockers). Further studies are first required to elucidate the mechanisms involved in ETS induced bronchoconstrictive reactions.

Acknowledgements

Project was generously funded by a contract from the Center for Indoor Air Research and by the NIH as a part of a specialized research center (NIH-NIAD AI 13401). The authors acknowledge the assistance of Marjorie McCants as study coordinator and Mrs. Pat Constant for word processing.

REFERENCES

Barnes, P.J., Baraniuk, J.N., and Belvisi, M.G., State of the Art. Neuropeptides in the respiratory tract. Part 1, *Am. Rev. Respir. Dis.,* 144, 1187-1198, 1991a.

Barnes, P.J., Baraniuk, J.N., and Belvisi, M.G., State of the Art. Neuropeptides in the respiratory tract. Part II, *Am. Rev. Respir. Dis.,* 144, 1391-1399, 1991b.

Becker, C.G., Dubin, T., and Wiedmann, H.P., Hypersensitivity to tobacco antigen, *Proc. Nat. Acad. Sci. USA*, 73, 1712-1716, 1976.

Brownson, R.C., Jackson-Thompson, J., Wilderson, J.C., Davis, J.R., Owens, J.W., and Fisher, E.B. Jr., Demographic and socioeconomic differences in beliefs about the health effects of smoking, *Am. J. Public Health,* 82, 99-103, 1992.

Centers for Disease Control, Discomfort from environmental tobacco smoke among employees at worksites with minimal smoking restrictions - United States, 1988. *Morbidity and Mortality Weekly Report,* 41(20), 351-354, 1992.

Centers for Disease Control and Prevention, Cigarette smoking among adults - United States, 1994. *Morbidity and Mortality Weekly Report,* 45(27), 588-590, 1996.

Chilmonczyk, B.A., Salmon, L.M., Magathin, K.K., Neveux, L.M., Palmoaki, G.E., Knight, G.J., and Pulkkienen, A.J., Association between exposure to environmental tobacco smoke and exacerbations of asthma in children, *New Engl. J. Med.,* 328, 1665-1669, 1993.

Dahms, T.C., Bolin, J.F., and Slavin, R.G., Passive smoking: Effects on bronchial asthma, *Chest,* 80, 630-634, 1981.

Danuser, B., Weber, A., Hartmann, A.L., and Krueger, H., Effects of a bronchoprovocation challenge test with cigarette sidestream smoke on sensitive and healthy adults, *Chest,* 103, 35-58, 1993.

Evans, D., Levison, M.J., Feldman, C.H., Clark, N.M., Wasilewski, Y., Levin, B., Mellins, R.B., The impact of passive smoking on emergency room visits of urban children with asthma, *Am. Rev. Respir. Dis.,* 135, 567-572, 1987.

Harkavy, J., Tobacco allergy in cardiovascular disease: A review. *Ann. Allergy,* 26, 447-459, 1968.

Kraemer, M.J., Richardson, M.A., Weiss, N.S., Furukawa, C.T., Shapiro, G.G., Pierson, W.E., and Bieman, C.W., Risk factors for persistent middle-ear effusions. Otis media, catarrh, cigarette smoke exposure, and atopy, *JAMA,* 249, 1022-1025, 1983.

Lehrer, S.B., Menon, P., Simlote, P., Rando, R., Musmand, J., McCants, M., Hughes, J., Lopez, M., and Salvaggio, J., The acute effects of environmental tobacco smoke exposure on asthmatics: studies with a dynamic test chamber, *Indoor Air,* 7(4), 269-277, 1997.

Menon, M.D., Rando, R., and Lehrer, S.B., Skin test reactivity to tobacco leaf (TLE) and smoke extracts (TSE) in relationship to clinical tobacco smoke reactivity, in *Proceedings of the American Academy of Pediatrics*, New Orleans, 1991, Abstract.

Menon, P.K., Rando, R., Poovey, H., and Lehrer, S., Extra pulmonary responses to side stream environmental tobacco smoke (ETS) during ETS provocation inhalation challenges, in *Proceedings of the American Academy of Pediatrics*, San Francisco, 1992.

Menon, P.K., Stankus, R.P., McCants, M.L., Salvaggio, J.E., and Lehrer, S.B., Tobacco leaf reactivity in "smoke-sensitive" asthmatics, *JACI,* 79, 261, 1987, Abstract.

Menon, P.K., Stankus, R.P., Rando, R.J., and Lehrer, S.B., Inhalation challenge studies with environmental tobacco smoke, *JACI,* 81, 324, 1988.

Murray, A.B., and Morrison, B.J., The effect of cigarette smoke from the mother on bronchial responsiveness and severity of symptoms in children with asthma, *J Allergy Clin Immunol,* 77, 575-581, 1986.

Murray, A.B., and Morrison, B.J., Passive smoking and the seasonal difference of severity of asthma in children, *Chest,* 94, 701-708, 1988.

Musmand, J.J., McCants, M., Lopez, M., and Lehrer, S.B., Tobacco smoke allergy: Is it a fallacy? *Clin Res,* 40(4), 805A, 1992, Abstract.

Musmand, J.J, White, M., Lopez, M., Kaliner, M.A., and Lehrer, S.B., The nasal response to environmental tobacco smoke, *J Allergy Immunol,* 91, 194, 1993.

Musmand, J., White, M., Mosimann, B., Lopez, M., Kaliner, M., and Lehrer, S., Asthmatics' respiratory tract neuropeptide response to acute environmental tobacco smoke exposure, *Chest,* 104(2), 11S, 1993.

Peden, D.B., Mechanisms of pollution-induced airway disease: in vivo studies, *Allergy,* 52(38 Suppl), 37-44, 1997.

Rando, R.J., Menon, P.K., Poovey, H.G., and Lehrer, S.B., Assessment of multiple markers of environmental tobacco smoke (ETS) in controlled, steady-state atmospheres in a dynamic test chamber, *American Industrial Hygiene Association Journal,* 53, 699-704, 1992.

Rando, R.J., Simlote, P., Salvaggio, J.E., and Lehrer, S.B., Environmental tobacco smoke: measurement and health effects of involuntary smoking, in *Indoor Air Pollution and Health*, Bardana, E.J, and Montonaro, A. Eds., Marcel Dekker, Inc., New York, 1997.

Raphael, G.D., Meredith, S.D., Baraniuk, J.N., Druce, H.M., Banks, S.M., and Kaliner, M.A., Pathophysiology of Rhinitis. 1. Assessment of the sources of protein in methacholine-induced nasal secretions, *Am. Rev. Respir. Dis.* 138, 413-420, 1988.

Raphael, G.D., Meredith, S.D., Baraniuk, J.N., Druce, H.M., Banks, S.M., and Kaliner, M.A., The Pathophysiology of Rhinitis. 11. Assessment of the sources of protein in histamine induced nasal secretions, *Am. Rev. Respir. Dis.* 139, 791-800, 1989.

Shepard, R.J., Collins, R., and Silverman, F., Passive exposure of asthmatic subjects to cigarette smoke. *Environmental Research,* 20, 392-402, 1979.

Stankus, R.P., Menon, P.K., Rando, R.J., Glindmeyer, H., Salvaggio, J.E., and Lehrer, S.B., Cigarette smoke sensitive asthma: Challenge studies. *JACI,* 82, 331-338, 1988.

Taylor, G., Tobacco smoke allergy - Does it exist? in *Environmental tobacco smoke effects on the nonsmoker.* Rylander, K., Ed., University of Geneva. 1974, pp 50-55.

U.S. Dept. of Health and Human Services, Public Health Service: Healthy People 2000, national health promotion and disease prevention objectives (Publication No. PHS91-50212), 1991.

Wiedemann, H.P., Mahler, D.A., Loke, J., Virgulto, J.A., Snyder, P., and Matthay, R.A., Acute effects of passive smoking on lung function and airway reactivity in asthmatic subjects, *Chest* 89, 180-185, 1986.

Zussman, B.M., Tobacco sensitivity in the allergic population, *J. Asthma Res.,* 11, 159-167, 1974.

Chapter 9

AIRBORNE CONTAMINANTS AND ASTHMA: THE EFFECTS OF OZONE, NO_2, SO_2 ENDOTOXIN AND DIESEL PARTICULATES ON NORMAL & ASTHMATIC LUNGS

David B. Peden, MD and Brian Boehlecke, MD, MSPH

I. INTRODUCTION

Asthma incidence and severity have increased markedly over the past three decades (Anonymous, 1987; Crain et al, 1994; Weiss et al, 1993). Environmental factors are thought to contribute to this phenomenon. Allergen exposure is an important environmental factor, with substantial evidence demonstrating that indoor allergen exposure is strongly linked to development of asthma, especially in children (Platts-Mills et al, 1997; Rosenstreich et al, 1997). Allergens of particular concern include those from house dust mites, mammalian pets, cockroach and *Alternaria* (Platts-Mills et al, 1997; Rosenstreich et al, 1997). Other environmental factors also effect asthma including exposure to outdoor irritant air pollutants (ozone, oxides of nitrogen, sulfur dioxide and particulate matter) and indoor non-allergen air contaminants (environmental tobacco smoke, microbes and endotoxin) (Anonymous, 1987; Anonymous, 1996; Anto & Sunyer, 1995; Balmes, 1993; Bates & Hazucha, 1973; Bates & Sizto, 1983; Folinsbee, 1993; Horstman et al, 1984; Horstman et al, 1995; Koren et al, 1995; Kreit et al, 1989; Peden, 1997a).

It is unclear what effect non-allergen environmental factors have had on the increased incidence of asthma. Many have suggested that outdoor air pollutants contribute to asthma pathogenesis. However, asthma incidence continues to increase despite general improvement in outdoor air quality. Von Mutius et al recently compared incidence of asthma and atopy in children born in eastern and western Germany shortly before reunification (von Mutius et al, 1994). Problems of air quality in eastern Germany, characterized by elevated levels of total suspended particulates and sulfur dioxide, were different from those observed in western Germany, which had higher ozone levels, but was otherwise less polluted. Interestingly, asthma and atopy were more frequent in children born and raised in the West while bronchitis was more prevalent in the East. Thus, the role of pollutant exposure in asthma pathogenesis is unclear and may be minimal.

On the other hand, exacerbation of asthma by air pollutants has been repeatedly demonstrated in epidemiological studies. Increases in ambient air ozone have been linked to increased emergency room visits, decreased peak expiratory flow rates, and medication use by asthmatics (Anonymous, 1996; Balmes, 1993; Bates & Sizto, 1983; Folinsbee, 1993; Peters et al, 1997;

1-56670-373-5/99/$0.00+$.50

Peterson & Saxon, 1996; Pope, 1991; Pope et al, 1995; Schwartz et al, 1992a; Schwartz & Dockery, 1992b; Schwartz et al, 1993; Timonen & Pekkanen, 1997). Usually, the effects are noted on the day following the increase in ambient air ozone, suggesting that a delayed inflammatory effect, rather than an immediate bronchoconstrictive effect, accounts for these changes. Similar observations have been made with particulate matter. Thus, outdoor air pollutants appear to have a direct effect on asthma severity.

Because most people spend the majority of time indoors, the effect of indoor irritants on asthma may be more important than the effect of outdoor pollutants. One such irritant is gram negative bacterial endotoxin. Endotoxin exposure in occupational settings is clearly associated with acute reductions in lung function and respiratory symptoms (Borm et al, 1996; Castellan et al, 1987; Castellan et al, 1984; Christiani et al, 1993; Clapp et al, 1993; Kateman et al, 1990; Vogelzang et al, 1998). Recent observations by Michel et al also indicate that disease severity of mite-sensitive asthmatics correlates with levels of endotoxin in domestic house dust (Michel et al, 1996). A possible explanation is that endotoxin, a known stimulant of neutrophilic inflammation (Michel et al, 1997), exacerbates pre-existing airway inflammation and amplifies allergic responses to mite allergen.

This chapter will review studies examining known effects of pollutants on airway inflammation in asthmatics. We will then describe how strategies employed to study outdoor pollutants can be applied to examine the effect of endotoxin on airway function in asthma. Comparison of the effects of non-allergenic irritants such as ozone and endotoxin may suggest common mechanisms by which asthmatics respond to these agents.

II. OZONE

A. EFFECT OF OZONE ON AIRWAY INFLAMMATION IN ASTHMA

In normal subjects, ozone exposure results in influx of neutrophils to the airway at exposures ranging from 0.10 to 0.4 ppm (Anonymous, 1996; Devlin et al, 1994; Holtzman et al, 1983; Koren et al, 1989), and stimulates release of IL-6, IL-8, PGE_2, LTB_4, TXB_2, fibronectin and elastase. Ozone also causes immediate decreases in lung function, which are neurally mediated and do not correlate with measures of airway inflammation (Folinsbee, 1994; Krishna et al, 1997; Matsuse et al, 1991; McDonnell et al, 1983; McDonnell et al, 1991).

In persons with allergic rhinitis or asthma, ozone exposure appears to induce both eosinophil as well as polymorphonuclear leukocyte (PMN) influx to epithelial surfaces (Bascom et al, 1990; McBride et al, 1994; Peden et al, 1995). This has been observed in the nasal mucosa of both allergic rhinitics and mite-sensitive asthmatics (Bascom et al, 1990; Peden et al, 1995). In the latter study, ozone also caused increases in eosinophil cationic protein found in surface liquid, suggesting that eosinophils may have been activated (Peden et al, 1995). Examination of bronchial inflammation in asthma following ozone exposure has yielded somewhat conflicting results. Two studies comparing the effect of ozone in normal and asthmatic subjects indicate that while asthmatics have a greater PMN response than normal subjects, neither group has an eosinophilic response. In the first of these studies, 5 asthmatics had a greater PMN and IL-8 response to ozone when compared to the response of 5 normal subjects (Basha et al, 1994). The second study with a larger number of subjects (19 asthmatics and 20 normal subjects) confirmed the original observation (Scannel et al, 1996). Broncho-alveolar lavage (BAL) fluid collected 18 hours after exposure to 0.2 ppm ozone for 2 hours revealed that asthmatics have increased levels of PMNs, LDH, total protein, myeloperoxidase, fibronectin and IL-8 following ozone exposure and that the asthmatics had a greater PMN response than non-asthmatics, with a trend for increased IL-8 levels as well (Scannel et al, 1996). Notably absent was any increase in eosinophils in the BAL fluid.

In contrast, following exposure to 0.16 ppm ozone for 8 hours, eight mite-sensitive asthmatics had a substantial increase in both PMNs and eosinophils in BAL fluid when compared to samples collected following exposure to clean filtered air. Bronchial samplings included traditional BAL fluid as well as separate examination of the first 20 ml recovered (the bronchial fraction), with the bronchial fraction yielding more striking differences between air and ozone exposure in asthmatics (Peden et al, 1997b). Consistent with this are preliminary data reported by Framptom and colleagues, in which a two-fold increase in eosinophils was observed in asthmatics following ozone exposure (Frampton et al, 1997). Thus, in these two groups of asthmatics, ozone induced both neutrophil and eosinophil influx to the airway. Collectively, these studies indicate that asthmatics react differently to ozone than normal subjects.

B. EFFECT OF OZONE ON AIRWAY RESPONSES TO ALLERGEN

Because only persons with asthma who are sensitized to an allergen are immediately affected by inhalation of that allergen, any action of a pollutant on IgE-mediated responses to allergen exposure on airway function represents a unique risk to this segment of the population from that pollutant. Controlled exposure studies have clearly demonstrated an effect of ozone on both immediate and late phase responses to inhaled allergen in atopic subjects.

The first report that ozone can alter immediate responses to inhaled allergen was made by Molfino et al, who observed increased sensitivity to inhaled allergen in seven grass sensitive subjects challenged immediately after exposure to 0.12 ppm ozone for one hour at rest (Molfino et al, 1991; Molfino, 1992). While a second study of 12 subjects employing this exposure protocol did not confirm this observation, others demonstrated that exposure to 0.25 ppm ozone for 3 hours with moderate exercise does enhance bronchial reactivity of both allergic asthmatics and allergic rhinitics to inhaled allergen (Ball et al, 1996; Jorres et al, 1996). Preliminary data from our group also shows that exposure to 0.16 ppm ozone for 8 hours with moderate exercise increased sensitivity of mite-sensitive asthmatics to inhaled mite allergen given 18 hours after the exposure.

Late phase response to allergen is also altered by ozone. In nasal studies of allergic asthmatics, 0.4 ppm ozone was found to enhance allergen-induced eosinophil influx, eosinophilic cationic protein (ECP) levels and IL-8 levels (Peden et al, 1995). A study in which exposure to 0.2 ppm ozone for 3 hours was followed by inhaled allergen challenge did not show any effect of allergen-associated levels of eosinophils, neutrophils, ECP, IL-8 recovered in nasal fluid. Taken together, these studies suggest that the effect of ozone on inhaled allergen challenge is dose related. Also, all of these studies were performed in mild asthmatics (Kreit et al, 1989). It is possible that persons with more severe disease may be more sensitive to the effect of ozone on both immediate and late phase responses to allergen.

III. NITROGEN DIOXIDE AND ASTHMA

Nitrogen dioxide has not been shown to have a significant direct acute effect on lung function in either asthmatic or non-asthmatic subjects (Anonymous, 1987). Examination of the effect of this gas on non-specific reactivity in asthmatics yields mixed results, with some investigators reporting enhanced bronchial responses to cholinergic or cold air challenge and others finding that NO_2 has no effect on airway responsiveness (Anonymous, 1996; Bauer et al, 1986). NO_2 may have an effect on airway inflammation, as normal subjects were found to have increased numbers of airway PMNs after exposure to this gas (Sandstrom et al, 1991). Overall, these data suggest that NO_2 , while not as potent as ozone, increases airway inflammation.

In a recent study of the effect of exposure to NO_2 on allergen-induced nasal inflammation in asthmatics, NO_2 alone was not associated with mast cell degranulation, eosinophil activation, neutrophil activation or increases in IL-8. However, exposure to NO_2 did enhance levels of ECP in nasal lavage fluid following allergen challenge, indicating an enhancement of the effect of allergen on eosinophil function (Wang et al, 1995). NO_2 has also been shown to have an effect on allergen-induced bronchospasm (Strand et al, 1997; Venaille et al, 1989). Tunnicliffe et al have reported that one hour's exposure to 0.4 ppm NO_2 enhanced immediate responsiveness to a fixed dose of allergen (Tunnicliffe et al, 1994). A combined exposure to 0.2 ppm SO_2 and 0.4 NO_2 for 6 hours also enhances immediate sensitivity of mild asthmatics to inhaled allergen (Peden et al, 1997a; Rylander et al, 1989). Exposure to NO_2 has also been reported to enhance late phase responses of asthmatics to inhaled allergen (Strand et al, 1997). Taken together, these studies demontrate that NO_2 is able to alter both allergic and nonspecific airway responses in asthmatics.

IV. DIESEL EXHAUST PARTICLES

Increases in ambient air particulate matter concentration have been associated with a number of health effects in humans, including increased need for bronchodilators, increased cough, increased episodes of bronchitis and even death (Anonymous, 1996; Pope, 1991; Pope et al, 1995; Schwartz & Dockery, 1992a; Schwartz & Dockery, 1992b; Schwartz et al, 1993). The mechanisms by which particulates specifically effect asthma are unclear. One way in which particles could influence asthma would be to act as non specific irritants on a background of eosinophilic inflammation. Alternatively, particulates could directly enhance TH-2 type inflammation via stimulation of pro-allergic cytokines, which in turn could promote a number of features of allergic inflammation, including IgE production and mast cell and eosinophil influx to the airway. Numerous animal and *in vitro* studies indicate that diesel exhaust particles (DEP) may well enhance allergic inflammation and airway hyperresponsiveness by alteration of TH2 inflammation (Muranaka et al, 1986; Peterson et al, 1996; Takafuji et al, 1987; Takafuji et al, 1989a; Takafuji, 1989b; Takano et al, 1997).

In humans, Diaz-Sanchez, Saxon and colleagues demonstrated that nasal challenge with 0.3 mg of DEP yields increased nasal IgE production within 4 days without any effect on IgG, IgA or IgM production in both atopic and non-atopic asthmatics (Diaz-Sanchez et al, 1994). These findings were dose specific with no effect being observed with 0.1 or 1.0 mg of DEP. Subsequent studies by this group have shown that DEP causes increased cytokine production in cells recovered from nasal lavage fluid. Cells recovered before challenge were generally found to have detectable mRNA levels for only γ-interferon, IL-2 and IL-13. In contrast, DEP challenge yielded cells which, in addition to message for γ-interferon, IL-2 and IL-13 had detectable levels of IL-4, IL-5, IL-6, and IL-10 mRNA. IL-4 protein was also measured in post challenge lavage (Diaz-Sanchez et al, 1997). While it is unclear which type of cells were present in lavage fluid before or after challenge, DEPs do appear to enhance TH-2 type inflammation.

The effect of coupling DEP with ragweed allergen was compared with nasal challenge with ragweed alone on allergen specific IgE and IgG responses in ragweed sensitive subjects (Diaz-Sanchez et al, 1997). DEP plus allergen resulted in a significant increase in allergen specific IgE and IgG_4 without an increase in total IgE and IgG. Likewise, DEP and allergen, when compared to DEP alone, was associated with increased expression of IL-4, IL-5, IL-6, IL-10, IL-13 and decreased expression of γ-interferon and IL-2.

Compared to other pollutants, such as ozone or metal ions, DEP appears to be unique in its effect on IgE production. The *in vivo* effect of DEP on IgE isotype switch can be replicated in vitro in isolated B lymphocytes with extracts from DEP which contain the polyaromatic hydrocarbon (PAH) fraction from these particles, as well as the specific PAH compounds phenanthrene and 2,3,7,8 tetracholorodibenzo-p-dioxin (Peterson & Saxon, 1996; Takenaka et al, 1995). Thus, polyaromatic hydrocarbons, by their action on B cells, appear to play a central role in the effect of diesel exhaust on allergic inflammation.

Sandstrom and colleagues have recently presented preliminary information suggesting that chamber exposure to diluted diesel exhaust being generated by a standardized engine induces airway responses in non-asthmatics. Perhaps the most interesting observation by this group was an increase in mast cell numbers in BAL fluid associated with diesel exposure. Because, diesel exhaust is a very complicated mixture containing DEPs of varying sizes, NO_2, and CO, among other components, it is not possible to directly attribute the effects of diesel exhaust to one substrate. However, these observations are consistent with the nasal experiments cited above, indicating an effect of DEP on TH-2 inflammatory responses, even in non-atopic subjects.

V. ENDOTOXIN AND ASTHMA

Endotoxin refers to the lipopolysaccharide (LPS) complex and associated proteins found in the outer layer of the cell wall of gram-negative bacteria. Although the lipid component (lipid A) is responsible for most of the toxic effects, variability of the biologic responses to endotoxin exposures in natural settings may be due in part to influences from associated cell wall components, which differ among bacterial species.

Most medical attention has focused on the toxicity of endotoxins in reaching the bloodstream from endogenous gram-negative bacterial infections of the host or via contaminated parenteral products. However, little endotoxin is detectable in the blood after inhalational exposure. Local uptake in the respiratory system appears to account for the manifestations observed in persons exposed to airborne endotoxin. Occupational settings with potential for high levels of airborne endotoxin include animal confinement and the storage or processing of organic materials (cotton dust, corn dust) e.g. textile mills and grain elevators (Clark et al, 1983; Jacobs, 1989).

Endotoxin has been proposed as a major factor in the pathogenesis of both lung function decrements and increases in non-specific bronchial reactivity which occur in persons exposed to organic dusts in occupational settings (Jacobs, 1989). Acute lung function decrements were found to be associated with endotoxin exposure in swine confinement workers (Donham et al, 1989), animal feed workers (Smid et al, 1994), and volunteers exposed to cotton dust in an experimental cardroom (Castellan et al, 1984). Non-asthmatic volunteers selected to show at least a 5% fall in FEV1 to a screening exposure to cotton dust were exposed for 6 hours in a model card room to various cotton dusts with airborne endotoxin concentrations up to

779 ng/m3. The drops in FEV1 were correlated with the endotoxin concentration but not the dust concentration. A no effect threshold of 10 ng/m3 of endotoxin was estimated (Castellan et al, 1987). Similar studies of textile workers have resulted in an estimated threshold for an acute effect on ventilatory function to be 100 to 200 ng/m3 (Rylander et al, 1985; Rylander, 1987a). These studies indicate that the endotoxin content of organic dusts plays a significant role in the pathogenesis of the acute airways obstruction which occurs in some persons when these dusts are inhaled.

Bioaerosols including gram-negative bacteria are also common contaminants of indoor air in residences and offices (Burge, 1995). Indoor air concentrations of microbial agents can be increased by humidifiers, ice machines, flush toilets, air conditioners, and evaporative humidification systems (Baur et al, 1988; McCunney, 1987; Morey et al, 1984; Rylander & Haglind, 1984; Spengler & Sexton, 1983). Because large volumes of water are aerosolized by office humidification systems, even minor contamination of the water can result in substantial concentrations of bacteria in the indoor air. Therefore, office workers may be exposed to higher levels of airborne gram-negative bacteria at work than at home or during usual outdoor activities. Although these organisms are susceptible to being rendered non-viable by aerosolization and drying, the endotoxin component of the cell wall resists degradation and retains its ability to exert a toxic effect on the respiratory system after inhalation.

Exposure to endotoxin, has been implicated as a factor modifying the clinical severity of asthma (Michel et al, 1994). The concentration of endotoxin in domestic house dust has been found to correlate with increased symptoms, increased medication use, and worsening lung function in mite allergic asthmatics (Michel et al, 1991a). This relationship was independent of the concentration of mite antigen in the dust. The authors suggested that the endotoxin might act synergistically with specific allergens or have a direct effect through its pro-inflammatory properties.

Recent surveys showed that both physician diagnosed asthma and wheezing without a specific diagnosis in primary school students were significantly associated with water damage to the home reported by parents (Maier et al, 1995; Williamson et al, 1997). The association with water damage was as strong as that for exposure to environmental tobacco smoke. No association was found with reported mold. Water damaged carpets and walls can lead to increased levels of bioaerosols including endotoxin in indoor air. Although no direct measurements of airborne fungi or endotoxin in the homes were made, the survey findings are consistent with the hypothesis that increased levels of bioaerosols including airborne endotoxin in indoor air may contribute to exacerbations of asthma.

Investigations of work-related upper respiratory irritation and other complaints labeled the "sick building syndrome" have found inconsistent relationships with airborne concentrations of microbial organisms, but better correlations were noted when examined for groups of buildings with the same type of ventilation (natural, mechanical, fully air-conditioned) (Harrison et al, 1992). In some instances airborne concentrations of the lipid moiety of endotoxin (LPS) have correlated with symptoms and others have suspected a mechanism involving endotoxins (Burge et al, 1987). In a recent prospective study Teeuw and co-workers found airborne endotoxin levels 6 to 7 fold higher in office buildings in which workers had a symptom prevalence above 15% compared to buildings in which symptoms were less frequent (Teeuw et al, 1994). More importantly, the airborne concentrations of endotoxin in these buildings of 100 to 408 ng/m3 (mean 254 ng/m3) exceed the levels found to be associated with significant falls in lung function in susceptible but non-asthmatic individuals or textile workers during 4 to 6 hour exposures to endotoxin contained in cotton dust in an experimental setting (Castellan et al, 1987; Rylander et al, 1985). Therefore the concentration of airborne microbial organisms and endotoxin in the indoor air in these buildings may be sufficiently high to pose a risk for adverse effects on both the upper and lower respiratory tract of susceptible individuals including asthmatics and persons with chronic bronchitis.

A. AIRWAY RESPONSES TO ENDOTOXIN (LPS) IN NORMALS

LPS inhalation causes a primarily neutrophilic response in the respiratory tract and is thought to be mediated by IL-1β and TNFα secretion, from alveolar macrophages (AMs) in the airway. IL-6, IL-8 and GM-CSF are also secreted by AMs and serum treated epithelial cells following LPS exposure (Khair et al, 1995). Bronchial challenge studies confirm that LPS induces airway inflammation (Michel et al, 1997; Sandstrom et al, 1992; Sandstrom et al, 1994; Schwartz et al, 1995; Schwartz et al, 1994). In a study of eight normal subjects, inhalation of 100,000 ng of LPS resulted in a 100 fold increase in PMNs in BAL fluid compared to control challenge with 3 of these subjects reporting symptoms associated with the procedure (Sandstrom et al, 1992). Another study, comparing the action of corn dust extract and LPS in normal subjects and atopic subjects who did not have airway hyperreactivity reported that both cause similar increases in airway PMN, TNFα, IL-1β, IL-6 and IL-8 levels (Blaski et al, 1996). Nasal inflammation (as shown by PMNs and IL-8 in nasal lavage fluid) is also increased in persons occupationally exposed to LPS (Wang et al, 1997). Thus, inhaled LPS causes airway inflammation, which is typified by PMN influx and production of TNFα, IL-1β, IL-6 and IL-8. This inflammation is associated with differing levels of symptoms in normal subjects. Most experience some degree of bronchospasm, cough, or sputum production. Higher doses may lead to fever and toxic pneumonitis (Sandstrom et al, 1994).

An increase in non-specific bronchial reactivity after dust exposure has been demonstrated in textile workers (Haglind et al, 1983), flour mill workers (Taytard et al, 1988), and non-atopic non-smoking volunteers exposed to cotton dust in an experimental chamber (Boehlecke et al, 1986). In the latter study the non-specific reactivity was still elevated one day after the exposure but had returned to the baseline value by 4 days. An increase in bronchial responsiveness after the work shift in non-smoking textile workers was reported to correlate with airborne endotoxin levels in the work place but not with dust levels suggesting that the endotoxin was the cause of the change in reactivity (Rylander & Bergstrom, 1993). The same investigator gave non-atopic students an inhalation challenge with an aerosol of cell bound endotoxin at a concentration of 300,000 ng/ml and found increased bronchial reactivity 4 hours after exposure (Rylander et al, 1989). Inhalational exposure to endotoxin in rats causes a rapid influx of polymorphonuclear leukocytes (PMNs) into the airways and an increase in non-specific airway reactivity 1-2 hours later (Pauwels et al, 1990). An influx of PMNs into the airways also occurs in guinea pigs after inhalation of endotoxin (Venaille et al, 1989). Thus, there is evidence linking inhalation of endotoxin with airway inflammation and an increase in non-specific airway reactivity.

As mentioned above, endotoxin is known to affect alveolar macrophages causing release of lysosomal enzymes, chemotactic factors such as LTB4, tumor necrosis factor, platelet activating factor, and interleukins 1,6, and 8 (Adams & Czuprynski, 1990; Brown & Donaldson, 1996; Fuhlbrigge et al, 1987; Kotloff et al, 1990; Yoshimura et al, 1987). It also can activate complement through both the direct and alternate pathways (Vukajlovich et al, 1987). The ensuing influx of PMN's into the airway walls with possible release of additional toxic factors such as free radicals may be the basis for the increase in airway reactivity noted after endotoxin inhalation. However, a recent study in guinea pigs showed that pretreatment with capsaicin reduced the PMN influx into the airways after intratracheal instillation of endotoxin (Jarreau et al, 1994). Moreover, this pretreatment also prevented an increase in airway responsiveness to aerosolized histamine which occurred following endotoxin inhalation in the absence of capsaicin pretreatment. This suggests that capsaicin-sensitive nerves and neuropeptides may be involved in the induction of airway inflammation and hyperresponsiveness by endotoxin.

B. AIRWAY RESPONSES OF ATOPIC NON-ASTHMATICS AND ASTHMATICS TO LPS

Atopy has been proposed as a risk factor for the acute airway obstruction response to organic dusts containing endotoxin. Textile workers with positive skin tests for aeroallergens and a family history of atopy showed larger falls in FEV1 over the work shift than non-atopic workers (Jones et al, 1980). Non-asthmatic volunteers with positive skin tests to aeroallergens (Sepulveda et al, 1984) or a history of a physician confirmed diagnosis of hay fever, manifest by rhinitis and conjunctivitis (Jacobs et al, 1993) showed significantly greater falls in FEV1 than non-atopic subjects when exposed to cotton dust in a model cardroom. Airborne dust and endotoxin concentrations for the exposures ranged up to 1 mg/m3 and 1020 ng/m3, respectively. The difference between the magnitude of declines in FEV1 was small indicating that mild atopy by itself is a relatively weak risk factor. Baseline bronchial reactivity to methacholine was greater in the atopics than the non-atopics in the latter study and was significantly correlated with the magnitude of the fall in FEV1 during exposure. However, other studies have not shown definite associations between pre-exposure non-specific airway reactivity and lung function decline after exposure to endotoxin (Boehlecke 1992; Schachter et al, 1981). Thus mild atopy is likely to be associated with a small enhancement of the airways obstruction occurring after inhalation of endotoxin. Pre-existing non-specific bronchial hyperreactivity may be an important risk factor in some individuals.

Inhalation of aeorsolized endotoxin solutions by mild asthmatics has been shown to cause a transient fall in lung function and an increase in non-specific airway reactivity. An average reduction of approximately 6-7% in FEV1 occurred 30 minutes after the inhalation of an aerosolized dose of 22,200 ng of the lipid moiety of endotoxin (LPS) in a small group of mild asthmatics who were not taking systemic or inhaled steroids (Michel et al, 1989). At 5 hours after exposure the FEV1 had improved slightly but was still reduced and the airway responsiveness to histamine was increased. These responses were seen in both atopic and non-atopic asthmatics as judged by skin testing. Neither of these effects were noted in a comparison group of non-asthmatics. Most of the asthmatics were smokers which may have influenced the results. However, a subsequent study of 16 asthmatics, 8 of whom were atopic, showed that the bronchial obstructive response and increased histamine bronchoreactivity after endotoxin were correlated with baseline histamine responsiveness, but not with atopy or smoking (Michel et al, 1992b). A further investigation in mild asthmatics using an acute inhalation of an aerosolized dose of 20,000 ng of LPS showed that the response is likely to involve an inflammatory process in the airways (Michel et al, 1992a). After the LPS challenge a significant fall in FEV1 was present at 45 minutes which decreased by 3 hours, was still detectable at 5 hours, and was gone by 6 hours. Peripheral blood showed an increase in TNF-α at 60 minutes and an increase in neutrophils at 6 hours consistent with an inflammatory response. Histamine responsiveness, an indirect indication of airway inflammation, was also increased at 6 hours. The absence of an immediate fall in FEV1 was interpreted to support a non-IgE mediated mechanism for the airway response. This hypothesis is supported by the lack of any effect on immediate skin test responses when the test antigens are contaminated with LPS (Michel et al, 1991b). Thus, inhalation of aerosols of endotoxin by asthmatics promotes an increase in airway reactivity which may be mediated by a non-allergic inflammatory response.

Both peripheral blood mononuclear cells and alveolar macrophages (AMs) from asthmatics generate increased levels of cytokines compared to cells recovered from normal subjects. With LPS stimulation, AMs from asthmatics generated twice the level of GM-CSF, three times the level of TNFα, and four times the amount of IL-8 produced by AMs from normal subjects (Hallsworth et al, 1994). Additionally, allergic asthmatics have increased levels of soluble CD14 (sCD-14) and LPS-binding protein (LPS-BP) in the airway following acute allergen challenge (Dubin et al, 1996). CD-14 serves as a receptor for LPS, and LPS-BP complexes with LPS to form a better ligand for CD-14. Many cells express CD-14 on their membranes, including monocytes and macrophages. CD-14 bearing cells can also shed this molecule into the circulation, where it is available to bind to cells, such as airway epithelial cells, which have little or no CD-14, rendering them responsive to LPS.

The studies just described suggest that atopic and asthmatic individuals are more susceptible to the effects of LPS than are normals. There is however some conflicting evidence. The production of an inflammatory response in the lungs of non-asthmatics by LPS aerosol inhalation has also been demonstrated by bronchoalveolar lavage (Sandstrom et al, 1992). Neutrophils increased 100-fold in lavage fluid 3 hours after inhalation of 100,000 ng LPS dissolved in saline compared to a control day lavage. Although the investigators did not comment on the visual appearance of the airways, three of eight subjects reported irritative airway symptoms after the exposure. Comparison of the effect of LPS on lung function, PMN influx, and production of IL-1β, IL-6, IL-8 and TNFα in normal subjects and atopic subjects without airway hyperresponsiveness revealed no differences between these two groups, including eosinophil influx (Blaski et al, 1996). Endotoxin-contaminated allergen solutions, when bronchoscopically instilled into a segment of the lower airway, caused a marked neutrophilic inflammatory response which was observed on follow-up bronchoscopy 18 hours later (Hunt et al, 1994). This contrasts with more typical allergen-induced inflammation, characterized by influx of both neutrophils (PMNs) and eosinophils, which was observed when endotoxin-free allergen was used. However, review of the literature does not reveal comparative dose response studies of inflammatory responses to LPS of asthmatics and non-asthmatic subjects, so it is not clear if asthmatics have enhanced PMN responses to LPS as has been reported with ozone (Peden et al, 1997).

The findings discussed above indicate that inhalation of endotoxin can induce airway inflammation, airway narrowing, and an increase in non-specific airway reactivity in normal and asthmatic individuals. By analogy with the findings cited for ozone, endotoxin inhalation might also enhance specific airway responsiveness in allergic asthmatics. In partial support of this hypothesis is the finding that pre-treatment of ovalbumin (OA) sensitized guinea pigs with capsaicin prevents the increased airway responsiveness to acetylcholine induced by repeated OA inhalations but does not prevent the development of airway eosinophilia (Matsuse et al, 1991). This suggests that a mechanism involving neuropeptides might be involved in antigen-induced airway hyperresponsiveness. Since capsaicin pre-treatment has also been shown to prevent the enhancement of non-specific airway reactivity after inhalation of endotoxin in non-sensitized guinea pigs, a common mechanism may be involved. Enhancement of such a common mechanism might increase airway hyperresponsiveness after antigen challenge even if the allergic response itself is not directly altered. Thus, interaction of the effects of endotoxin and allergen through a common pathway might produce a synergistic effect on airway reactivity when both agents are encountered. This might account for the finding of an association between increased clinical symptoms in mite-sensitive asthmatics and the endotoxin content of the house dust in their residence (Michel et al, 1991a).

C. EOSINOPHILIC INFLAMMATION AND LPS

Allergic inflammation is modulated by a number of cytokines and is characterized by increased numbers of mast cells and eosinophils, although neutrophils may also play a role (Anonymous, 1987). Eosinophils appear to be the significant inflammatory effector cell in allergic inflammation, with eosinophil numbers correlating well with disease severity, bronchospasm and bronchial hyperreactiviy (Anonymous, 1987). Endotoxin is traditionally associated with neutrophil influx, rather than eosinophil recruitment. However, a number of recent observations in animals suggest that LPS may enhance eosinophil responses as well (Bozza, et al, 1994a; Bozza et al, 1994b; Henriques et al, 1996; Macari et al, 1996; Weg et al, 1995). Skin challenge of guinea pigs with LPS doses ranging from 50-1000 ng revealed a dose-dependent influx of eosinophils 4 hours after challenge which was reduced by inhibition of TNFα (Weg et al, 1995). LPS and ovalbumin (OVA) both induce pleural eosinophilia in OVA-sensitized mice. Interestingly, treatment with anti-IL-5 monoclonal antibodies significantly blunted the allergen-associated eosinophil influx, but not that associated with LPS, suggesting that another cytokine was important in LPS-induced eosinophilia in these animals (Bozza et al, 1994b). At present, there are few data indicating that LPS effects eosinophil action in humans, but this needs further study (Kita et al, 1995).

Overall, the available data suggest that asthmatics may be more sensitive to the effect of inhaled LPS than normal subjects. This may be due to a non-specific increased responsiveness to LPS in all persons with chronic airway inflammation, including asthmatics. However, a more specific effect may be present in asthmatics with enhancement of immediate and delayed inflammatory responses to allergen due to a "priming" effect of the LPS. As discussed above, several mechanisms for such an effect are possible and deserve further study.

VI. STRATEGIES FOR STUDYING AIRBORNE CONTAMINANTS USING ENDOTOXIN AS A MODEL

Various pollutants, including ozone, nitrogen dioxide and diesel exhaust, generally appear to have the following effects in asthmatics:

1. To directly enhance TH2-type inflammation in the airways;
2. To enhance immediate immunologically mediated responses to allergen;
3. To enhance the inflammation that follows allergen exposure of the airway.

Using similar strategies to those outlined for traditional pollutants, we have initiated studies of the effect of endotoxin on allergic inflammation. These studies involve controlled dose inhalation challenge studies of the effect of endotoxin exposure on subsequent immediate bronchial responses to inhaled allergen as well as studies in which asthmatics are challenged with LPS and then allergen instilled directly in the nose to determine if LPS enhances late phase inflammation. Also to be examined is the nasal and bronchial response to LPS after acute allergen challenge as well as comparing the responses of asthmatics and normal subjects to LPS.

Finally, it should be noted that genetic polymorphisms may influence susceptibility of a given individual to a particular pollutant. Animal data suggest that this may be true for ozone (Kleeberger et al, 1997). Likewise, TNFα polymorphisms are associated with differences in the *in vitro* response of monocytes to LPS as well as in chronic bronchitis (Derkx et al, 1995; Huang et al, 1997). Thus, gene-environment interactions may be important determinants of increased susceptibility to a given agent. Furthermore, if polymorphisms associated with increased susceptibility occur in persons with other illnesses, such as asthma, then the effect of these genes may be magnified. Thus, an additional strategy in examining the effect of LPS on human health is to develop a reliable phenotype profile (e.g. increased eosinophilia, airway hyper-responsiveness or GM-CSF following exposure to LPS in asthmatics) so that responder and non-responder groups can be identified and screened for candidate response genes.

REFERENCES

Adams JL, Czuprynski CJ. Bacterial LPS induces release of TNF_α from bovine, peripheral blood monocytes and alveolar macrophages *in vitro*. *J Leukocyte Biol*, 48:549-56, 1990.

Anonymous. Guidelines for the diagnosis and management of asthma. Expert Panel Report 2. *NIH Publication,* 97-4051, July, 1997.

Anonymous. Standardization of spirometry-1987 update. American Thoracic Society, *Am Rev Respir Dis*, 136:1285-1298, 1987.

Anonymous. Health effects of outdoor air pollution. Committee of the Environmental and Occupational Health Assembly of the American Thoracic Society. *Amer J Resp Crit Care Med,* 153(1):3-50, 1996.

Anto JM, Sunyer J. Nitrogen dioxide and allergic asthma: Starting to clarify an obscure association. *Lancet,* 345:402-403, 1995.

Ball BA, Folinsbee LJ, Peden DB, Kehrl HR. Allergen bronchoprovocation of patients with mild allergic asthma after ozone exposure. *J Allergy Clin Immunol*, 98(3):563-72, 1996.

Balmes JR. The Role of Ozone Exposure in the Epidemiology of Asthma. *Environ Health Perspect*, 101:219-224, 1993.

Bascom R, Naclerio RM, Fitzgerald TK, *et al.* Effect of ozone inhalation on the response to nasal challenge with antigen of allergic subjects. *Am Rev Respir Dis*, 142:594-601, 1990.

Basha MA, Gross KB, Gwizdala CJ, Haidar AH, Popovich J Jr. Bronchoalveolar lavage neutrophilia in asthmatic and healthy volunteers after controlled exposure to ozone and filtered purified air. *Chest*, 106(6):1757-65, 1994.

Bates DV, Hazucha MJ. The Short-Term Effects of Ozone on the Human Lung. *US Gov Print Off*, 93-15:507-540, 1973.

Bates DV, Sizto R. Relationship Between Air Pollutant Levels and Hospital Admissions in Southern Ontario. *Can J Public Health*, 74:117-122, 1983.

Bauer MA, Utell MJ, Morrow PE, *et al.* Inhalation of 0.30 ppm Nitrogen Dioxide Potentiates Exercise-Induced Bronchospasm in Asthmatics. *Am Rev Respir Dis*, 134:1203-1208, 1986.

Baur X, Behr J, Dewair M, Ehret W, Fruhmann G, Vogelmeier C, Weiss W, Zinkernagel V. Humidifier lung and humidifier fever. *Lung*, 166:113-124, 1988.

Blaski CA, Clapp WD, Thorne PS, Quinn TJ, Watt JL, Fress KL, Yagla SJ, Schwartz DA. The role of atopy in grain dust-induced airway disease. *Am J Resp Crit Care Med*, 154(2 Pt1):334-40, 1996.

Boehlecke B, Schreiber R, Warrenfelt J. Non-specific reactivity increased by exposure to cotton dust. *Abstracts of the Third International Conference on Environmental Lung Disease*, October 15-18, 1986, Montreal, Canada.

Boehlecke B. *Report to NIOSH on Cooperative Agreement #81-OH-158,* 1992.

Borm PJ, Schins RP, Derhaag TJ, Kant I, Jorna TH. Cross-shift changes in blood inflammatory markers occur in the absence of airway obstruction in workers exposed to grain dust. *Chest*, 109(4):1078-85, 1996.

Bozza PT, Castro-Faria-Neto HC, Penido C, Larangeria AP, das Gracas M, Henriques MO, Silva PM, Martins MA, dos Santos RR, Cordeiro RS. Requirement for lymphocytes and resident macrophages in LPS-induced pleural eosinophil accumulation. *J Leukocyte Biology,* 56:151-8, 1994a.

Bozza PT, Castro-Faria-Neto HC, Penido C, Larangeira AP, Silva PM, Martins MA, Cordeiro RS. IL-5 accounts for the mouse pleural eosinophil accumulation triggered by antigen but not by LPS. *Immunopharmacology*, 27(2):131-6, 1994b.

Brown DM, Donaldson K. Wool and grain dusts stimulate TNF secretion by alveolar macrophages *in vitro*. *Occupat & Environ Med,* 53(6):387-93, 1996.

Burge P, Hedge A, Wilson S, Bass J, Robertson A. Sick building syndrome: a study of 4373 office workers. *Ann Occup Hyg,* 31:493-504, 1987.

Burge H. Bioaerosols: prevalence and health effects in the indoor environment. *J Allergy Clin Immunol*, 86(5):687-701, 1990.

Burge HA. Aerobiology of the indoor environment. *Occupat Med,* 10(1):27-40, 1995.

Castellan RM, Olenchock SS, Hankinson JL, Millner PD, Cocke JB, Bragg CK, Perkins HH, Jacobs RR. Acute bronchoconstriction induced by cotton dust: dose-related response to endotoxin and other dust factors. *Ann Int Med,* 101:157-163 1984.

Castellan RM, Olenchock SA, Kinsley KB, Hankinson JL. Inhaled endotoxin and decreased spirometric values: an exposure-response for cotton dust. *N Engl J Med*, 317(10):605-610, 1987.

Christiani DC, Wegman DH, Eisen EA, Ye TT, Lu PL, Olenchock SA. Cotton dust and gram-negative bacterial endotoxin correlations in two cotton textile mills. *Amer J Ind Med*, 23(2):333-42,1993.

Clapp WD, Thorne PS, Frees KL, Zhang X, Lux CR, Schwartz DA. The effects of inhalation of grain dust extract and endotoxin on upper and lower airways. *Chest*, 104(3):825-30,1993.

Clark S, Rylander R, Larsson L. Airborne bacteria, endotoxin and fungi in dust in poultry and swine confinement buildings. *Amer Ind Hyg Assoc J,* 44(7):537-41, 1983.

Crain EF, Weiss KB, Bijur PE, Hersh M, Westbrook L, Stein RE. An estimate of the prevalence of asthma and wheezing among inner-city children. *Pediatrics*, 94(3):356-62, 1994.

Derkx HHF, Bruin KF, Jongeneel CV, deWaal LP, Brinkman BMN, Verweij CL, Houwing-Duistermaat JJ, Rosendal FR, vanDeventer SJH. Familial differences in endotoxin-induced TNF release in whole blood and peripheral blood mononuclear cells *in vitro*; relationship to TNF gene polymorphism. *J Endotox Res*, 2:19-25, 1995.

Devlin RB, McKinnon KP, Noah T, Becker S, Koren HS. Ozone-induced release of cytokines and fibronectin by alveolar macrophages and airway

epithelial cells. *Amer J Physiol,* 266(6 Pt 1):L612-9,1994.
Diaz-Sanchez D, Dotson AR, Takenaka H, Saxon A. Diesel exhaust particles induce local IgE production in vivo and alter the pattern of IgE messenger RNA isoforms. *J Clin Invest,* 94:1417-1425, 1994.
Diaz-Sanchez D, Tsien A, Fleming J, Saxon A. Combined diesel exhaust particulate and ragweed allergen challenge markedly enhances human *in vivo* nasal ragweed-specific IgE and skews cytokine production to a T helper cell 2-type pattern. *J Immunol,* 158(5):2406-13, 1997.
Donham K, Haglind P, Peterson Y, Rylander R, Bolin L. Environmental and health studies of farm workers in Swedish swine confinement buildings. *Br J Ind Med,* 46:31-37, 1989.
Dubin W, Martin TR, Swoveland P, Leturcq DJ, Moriarty AM, Tobias PS, Bleecker ER, Goldblum SE, Hasday JD. Asthma and endotoxin: lipopolysaccharide-bindingprotein and soluble CD14 in bronchoalveolar compartment. *Amer J Physiol,* 270(5 Pt 1):L736-44, 1996.
Folinsbee LJ. *Human Health Effects of Air Pollution. Environ Health Perspect,* 100:45-56, 1993.
Folinsbee LJ, Horstman DH, Kehrl HR, *et al.* Respiratory responses to repeated prolonged exposure to 0.12 ppm ozone. *Am J Respir Crit Care Med,* 149:98-105, 1994.
Frampton MW, Balmes JR, Cox C, KreinPM, Speers DM, Tsai Y, Utell MJ. Mediators of inflammation in bronchoaleolar lavage fluid from nonsmokers, smokers and asthmatics subjects exposed to ozone-a collaborative study. *Health Effects Institute Research Report,* 78: 73-79 June.
Fuhlbrigge RC, Chaplin DD, Kiely JM, Unanue ER. Regulation of interleukin 1 gene expression by adherence and lipopolysaccharide. *J Immunol,* 138:3799-802, 1987.
Haglind P, Bake B, Belin L. Is mild byssinosis associated with small airways disease? *Eur J Respir Dis,* 64:449-59, 1983.
Hallsworth MP, Soh CP, Lane SJ, Arm JP, Lee TH. Selective enhancement of GM-CSF, TNF-alpha, IL-1 beta and IL-8 production by monocytes and macrophages of asthmatic subjects. *Eur Resp J,* 7(6):1096-102, 1994.
Harrison J, Pickering CAC, Faragher EB, Austwick PKC, Little SA, Lawton L. An investigation of the relationship between microbial and particulate indoor air pollution and the sick building syndrome. *Respir Med,* 86:225-235, 1992.
Henriques GM, Miotla JM, Cordeiro SB, Wolitsky BA, Woolley ST, Hellewell PG. Selectins mediate eosinophil recruitment in vivo: a comparison with their role in neutrophil influx. *Blood,* 87:5297-304, 1996.
Holtzman MJ, Fabbri LM, O'Byrne PM, *et al.* Importance of Airway-Inflammation for Hyperresponsiveness Induced by Ozone. *Am Rev*

Respir Dis, 127:686-690, 1983.

Horstman D, Roger LJ, McDonnell W, *et al.* Increased specific airway resistance (SRaw) in asthmatics exercising while exposed to 0.18 ppm ozone (O3). *The Physilogist,* 27:212 1984.

Horstman DH, Ball BA, Brown J, Gerrity T, Folinsbee LJ. Comparison of pulmonary responses of asthmatic and nonasthmatic subjects performing light exercise while exposed to a low level of ozone. *Tox Indust Health*, 11(4):369-85, 1995.

Huang SL, Su CH, Chang SC. Tumor necrosis factor-alpha gene polymorphism in chronic bronchitis. *Am J Resp Crit Care Med*, 156(5):1436-9,1997.

Hunt LW, Gleich GJ, Ohnishi T, Weiler DA, Mansfield ES, Kita H, Sur S. Endotoxin contamination causes neutrophilia following pulmonary allergen challenge. *Am J Resp Crit Care Med*, 149(6):1471-5, 1994.

Jacobs RR. Airborne endotoxins: an association with occupational lung disease. *Appl Ind Hyg,* 4:50-56, 1989.

Jacobs RR, Boehlecke B, von Hage-Hamsten M, Rylander R. Bronchial reactivity, atopy, and airway response to cotton dust. *Am Rev Respir Dis*, 148:19-24, 1993.

Jarreau P-H, D'Ortho M-P, Boyer V, Harf A, Macquin-Mavier I. Effects of capsaicin on the airway responses to inhaled endotoxin in the guinea pig. *Am J Respir Cri. Care Med,* 149:128-33, 1994.

Jones RN, Butcher BT, Hammad YY, Diem JE, Glindmeyer HW, Lehrer SB, Hughes JM, Weill H. Interaction of atopy and exposure to cotton dust in the bronchoconstrictor response. *Br J Ind Med,* 37:141-6. 1980.

Jorres R, Nowak D, Magnussen H. The effect of ozone exposure on allergen responsiveness in subjects with asthma or rhinitis. *Amer J Resp Crit Care Med,* 153(1):56-64, 1996.

Kateman E, Heederik D, Pal TM, Smeets M, Smid T, Spitteler M. Relationship of airborne microorganisms with the lung function and leucocyte levels of workers with a history of humidifier fever. *Scand J Work, Environ & Health,* 16(6):428-33, 1990.

Khair OA, Devalia JL, Abdelaziz MM, Sapsford RJ, Davies RJ. Effect of erythromycin on Haemophilus influenzae endotoxin-induced release of IL-6, IL-8 and sICAM-1 by cultured human bronchial epithelial cells. *Eur Resp J,* 8(9):1451-7, 1995.

Kita H, Mayeno AN, Weyand CM, Goronzy JJ, Weiler DA, Lundy SK, Abrams JS, Gleich GJ. Eosinophil-active cytokine from mononuclear cells cultured with L-tryptophan products:an unexpected consequence of endotoxin contamination. *J Allergy Clin Immunol,* 95:1261-7, 1995.

Kleeberger SR, Levitt RC, Zhang LY, Longphre M, Harkema J, Jedlicka A, Eleff SM, DiSilvestre D, Holroyd KJ. Linkage analysis of susceptibility to ozone-induced lung inflammation in inbred mice. *Natюre Genetics,* 17(4):475-8, 1997.

Koren HS, Devlin RB, Graham DE, Mann R, McGee MP, Horstman DH, Kozumbo WJ, Becker S, House DE, McDonnell WF, *et al.* Ozone-induced inflammation in the lower airways of human subjects. *Amer Rev Resp Dis,* 139(2):407-15, 1989.

Koren HS, Bromberg PA. Respiratory responses of asthmatics to ozone. *Int Arch Allergy Immunol,* 107(1-3):236-8, 1995.

Kotloff RM, Little J, Elias JA. Human alveolar macrophage and blood monocyte interleukin-6 production. *Am J Respir Cell Mol Biol,* 3:497-505, 1990.

Kreit JW, Gross KB, Moore TB, Lorenzen TJ, D'Arcy J, Eschenbacher WL. Ozone-induced changes in pulmonaryfunction and bronchial responsiveness in asthmatics. *J Appl Physiol,* 66(1):217-22, 1989.

Krishna MT, Springall D, Meng QH, Withers N, Macleod D, Biscione G, Frew A, Polak J, Holgate S. Effects of ozone on epithelium and sensory nerves in the bronchial mucosa of healthy humans. *Amer J Resp Crit Care Med,* 156(3 Pt 1):943-50, 1997.

Macari DM, Teixeria MM, Hellewell PG. Priming of eosinophil recruitment in vivo by LPS pretreatment. *J Immunol,* 157:1684-92, 1996.

Maier WC, Arrighi HM, Redding GJ, Morray BH, Llewllyn CE. Indoor risk factor for asthma and wheezing in Seattle primary school children. *Am J Respir Crit Care Med*, 151:A31, 1995.

Matsuse T, Thomson RJ, Chen X-R, Salari H, Schellenberg RR. Capsaicin inhibits airway hyperresponsiveness but not lipoxygenase activity or eosinophilia after repeated aerosolized antigen in guinea pigs. *Am Rev Respir Dis,* 144:368-372, 1991.

McBride DE, Koenig JQ, Luchtel DL, Williams PV, Henderson WR Jr. Inflammatory effects of ozone in the upper airways of subjects with asthma. *Amer J Resp Crit Care Med,* 149(5):1192-7, 1994.

McCunney RJ. The role of building construction and ventilation in indoor air pollution. Review of a recurring problem. *NY State J Med,* 87:203-209, 1987.

McDonnell WF, Horstman DH, Hazucha MJ, *et al.* Pulmonary Effects of Ozone Exposure During Exercise: Dose-Response Characteristics. *J Appl Physiol Respirat Environ Exercise Physiol,* 54(5):1345-1352, 1983.

McDonnell WF, Kehrl HR, Abdul Salaam S, *et al.* Respiratory response of humans exposed to low levels of ozone for 6.6 hours. *Arch Environ Health*, 46:145-150, 1991.

Michel O, Duchateau J, Sergysels R. Effect of inhaled endotoxin on bronchial reactivity in asthmatic and normal subjects. *J Appl Physiol ,* 66:1059-1064, 1989.

Michel O, Ginanni R, Duchateau J, et. al. Domestic endotoxin exposure and clinical severity of asthma. *Clin Exp Allergy,* 21:441-448, 1991a.

Michel O, Ginanni R, LeBon B, Duchateau J. Effect of endotoxin contamination on the antigenic skin test response. Ann Allergy, 66:39-

42, 1991b.

Michel O, Ginanni R, Le Bon B, Content J, Duchateau J, Sergysels R. Inflammatory response to acute inhalation of endotoxin in asthmatic patients. *Amer Rev Resp Dis,* 146(2):352-7,1992a.

Michel O, Ginanni R, Sergysels R. Relation between the bronchial obstructive response to inhaled lipopolysaccharide and bronchial responsiveness to histamine. *Thorax,* 47:288-291, 1992b.

Michel O, Duchateau J, Sergysels R. Are endotoxins an etiopathogenic factor in asthma? *Am J Ind Med,* 25:129-130, 1994.

Michel O, Kips J, Duchateau J, Vertongen F, Robert L, Collet H, Pauwels R, Sergysels R. Severity of asthma is related to endotoxin in house dust. *Amer J Resp Crit Care Med,* 154(6 Pt1):1641-6, 1996.

Michel O, Nagy AM, Schroeven M, Duchateau J, Neve J, Fondu P, Sergysels R. Dose-response relationship to inhaled endotoxin in normal subjects. *Amer J Resp Critical Care Med,* 156:1157-11645, 1997.

Molfino NA, Wright SC, Katz I, Tarlo S, Silverman F, McClean PDA, Szalai JP, Raizenne M, Slutsky AS, Zamel N. Effect of low concentrations of ozone on inhaled allergen responses in asthmatic subjects. *Lancet,* 338:191-203, 1991.

Molfino NA. Effect of Air Pollution on Atopic Asthma. *Medicina,* 52:363-367, 1992.

Morey PR, Hodgson MJ, Sorenson WG, Kullman GJ, Rhodes WW, Visvesvara GS. Environmental studies in moldy office buildings: biological agents, sources and preventative measures. *Ann Am Conf Gov Ind Hyg,* 10:21-35, 1984.

Muranaka M, Suzuki S, Koizumi K, Takafuji S, Miyamoto R, Ikemori R, Tokiwa H. Adjuvant activity of diesel exhaust particulates for the production of IgE antibody in mice. *J Allergy Clin Immunol*, 77:616-623, 1986.

Pauwels RA, Kips JC, Peleman RA, Vanderstraeten ME. The effect of endotoxin inhalation on airway responsiveness and cellular influx in rats. *Am Rev Respir Dis,* 141:540-5, 1990.

Peden DB, Setzer RW Jr, Devlin RB. Ozone exposure has both a priming effect on allergen-induced responses and an intrinsic inflammatory action in the nasal airways of perennially allergic asthmatics. *Amer J Resp Crit Care Med,* 151(5):1336-45, 1995.

Peden DB. Mechanisms of pollution-induced airway disease: *in vivo* studies. *Allergy,* 52(38 Suppl):37-44; discussion 57-8, 1997a.

Peden DB, Boehlecke BB, Horstman D, Devlin R. Prolonged, acute exposure to 0.16 ppm ozone induces eosinophilic airway inflammation in allergic asthmatics. *J Allergy Clin Immunol*, 100:802-808 1997b.

Peters A, Dockery DW, Heinrich J,Wichmann HE. Short-term effects of particulate air pollution on respiratory morbidity in asthmatic children. *Eur Resp J,* 10(4):872-9, 1997.

Peterson B, Saxon A. Global increases in allergic respiratory disease: the possible role of diesel exhaust particles. *Ann Allergy, Asthma, & Immunol*, 77(4):263-8; 1996.

Platts-Mills TA, Sporik RB, Chapman MD, Heymann PW. The role of domestic allergens. *Ciba Foundation Symposium*, 206:173-85, 1997.

Pope CA. Respiratory illness associated with community air pollution in Utah, Salt Lake and Cache Valleys. *Arch Environ Hlth*, 46:90-97, 1991.

Pope CA, Dockery DW, Schwartz J. Review of epidemiological evidence of health effects of particulate air pollution. *Inhal Toxicol,* 7:1-18, 1995.

Rosenstreich DL, Eggleston P, Kattan M, Baker D, Slavin RG, Gergen P, Mitchell H, McNiff-Mortimer K, Lynn H, Ownby D, Malveaux F. The role of cockroach allergy and exposure to cockroach allergen in causing morbidity among inner-city children with asthma. *N Engl J Med,* 336(19):1356-63, 1997.

Rusznak C, Devalia JL, Davies RJ. Airway response of asthmatic subjects to inhaled allergen after exposure to pollutants. *Thorax*, 51(11):1105-8, 1996.

Rylander R, Haglind P. Airborne endotoxins and humidifier disease. *Clin Allergy,* 14:109-112 1984.

Rylander R, Haglind P, Lundholm M. Endotoxin in cotton dust and respiratory function decrement among cotton workers in an experimental cardroom. *Am Rev Respir Dis,* 131:209-213, 1985.

Rylander R. Role of endotoxins in the pathogenesis of respiratory disorders. *Eur J Resp Dis - Sup,* 154:136-44, 1987a.

Rylander R. The role of endotoxin for reactions after exposure to cotton dust. *Amer J Ind Med,* 12(6):687-97, 1987b.

Rylander R, Bake B, Fischer JJ, Helander IM. Pulmonary function and symptoms after inhalation of endotoxin. *Am Rev Respir Dis*, 140:981-986, 1989.

Rylander R, Bergstrom R. Bronchial reactivity among cotton workers in relation to dust and endotoxin exposure. *Ann Occupat Hyg*, 37(1):57-63, 1993.

Sandstrom T, Stjernberg N, Eklund A, *et al.* Inflammatory Cell Response in Bronchoalveolar Lavage Fluid After Nitrogen Dioxide Exposure of Healthy Subjects - A Dose-Response Study. *Eur Respir J,* 4:332-333, 1991.

Sandstrom T, Bjermer L, Rylander R. Lipopolysaccharide (LPS) inhalation in healthy subjects increases neutrophils, lymphocytes and fibronectin levels in bronchoalveolar lavage fluid. *Eur Respir J,* 5:992-996, 1992.

Sandstrom T, Bjermer L, Rylander R. Lipopolysaccharide (LPS) inhalation in healthy subjects causes bronchoalveolar neutrophilia, lymphocytosis, and fibronectin increase. *Amer J Ind Med,* 25(1):103-4, 1994.

Scannell C, Chen L, Aris RM, Tager I, Christian D, Ferrando R, Welch B,

Kelly T, Balmes JR. Greater ozone-induced inflammatory responses in subjects with asthma. *Amer J Resp Crit Care Med,* 154(1):24-9, 1996.

Schachter EN, Brown S, Zuskin E, Buck M, Kolack B, Bouhuys A. Airway reactivity in cotton-bract-induced bronchospasm. *Am Rev Respir Dis,* 123:273-6, 1981.

Schwartz DA, Thorne PS, Jagielo PJ, White GE, Bleuer SA, Frees KL. Endotoxin responsiveness and grain dust-induced inflammation in the lower respiratory tract. *Amer J Physiol,* 267(5 Pt1):L609-17, 1994.

Schwartz DA, Thorne PS, Yagla SJ, Burmeister LF, Olenchock SA, Watt JL, Quinn TJ. The role of endotoxin in grain dust-induced lung disease. *Amer J Resp Crit Care Med,* 152(2):603-8,1995.

Schwartz J, Dockery D. Particulate air pollution and daily mortality in Stubenville, Ohio. *Am J Epidemiol,* 135:12-19, 1992a.

Schwartz J, Dockery DW. Increased Mortality in Philadelphia Associated with Daily Air Pollution Concentrations. *Am Rev Respir Dis*, 145:600-604, 1992b.

Schwartz J, Slater D, Larson TV, *et al.* Particulate air pollution and hospital emergency room visits for asthma in Seattle. *Am Rev Respir Dis,* 147:826-831, 1993.

Sepulveda MJ, Castellan RM, Hankinson JL, Cocke JB. Acute lung function response to cotton dust in atopic and nonatopic individuals. *Br J Ind Med*, 41:487-91, 1984.

Silverman F. Asthma and respiratory irritants (ozone). *Environ Hlth Perspec*, 29:131-6, 1979.

Smid T, Heederik D, Houba R, Quanjer PH. Dust- and endotoxin-related acute lung function changes and work-related symptoms in workers in the animal feed industry. *Am J Ind Med,* 25:877-888, 1994.

Spengler JD, Sexton K. Indoor air pollution: a public health perspective. *Science*, 221:9-17, 1983.

Strand V, Rak S, Svartengren M, Bylin G. Nitrogen dioxide exposure enhances asthmatic reaction to inhaled allergen in subjects with asthma. *Am J Resp Crit Care Med,* 155(3):881-7, 1997.

Takafuji S, Suzuki S, Koizumi K, Tadakoro K, Miyamoto T, Ikemori R, Muranaka M. Diesel-exhaust particulates inoculated by the intranasal route have an adjuvant activity for IgE production in mice. *J Allergy Clin Immunol,* 79:639-645, 1987.

Takafuji S, Suzuki S, Koizumi K, Tadakoro K, Ohashi H, Muranaka M, Miyamoto T. Enhancing effect of suspended particulate matter on the IgE antibody production in mice. *Int Arch Allergy Appl Immunol,* 90: 1-7, 1989a.

Takafuji S, Suzuki S, Muranaka M, Miyamoto T. Influence of environmental factors on IgE production, IgE, mast cells and the allergic response. *Ciba Found Symp,* 147:188-204, 1989b.

Takano H, Yoshikawa T, Ichinose T, Miyabara Y, Imaoka K, Sagai M.

Diesel exhaust particles enhance antigen-induced airway inflammation and local cytokine expression in mice. *Am J Resp Crit Care Med,* 156(1):36-42, 1997.

Takenaka H, Zhang K, Diaz-Sanchez D, Tsien A, Saxon A. Enhanced human IgE production results from exposure to the aromatic hydrocarbons from diesel exhaust particles: direct effects on B-cell IgE production. *J Allergy Clin Immunol,* 95:103-115, 1995.

Taytard A, Tessier JF, Faugers JG, Vergeret J, Frequr P. Respiratory function and bronchial reactivity in mill workers. *Eur J Epidem,* 4:326-330, 1988.

Teeuw KB, Vandenbroucke-Grauls CM, Verhoef J. Airborne gram-negative bacteria and endotoxin in sick building syndrome. A study in Dutch governmental office buildings. *Arch Int Med,* 154(20):2339-45, 1994.

Timonen KL, Pekkanen J. Air pollution and respiratory health among children with asthmatic or cough symptoms. *Am J Resp Crit Care Med,* 156(2 Pt 1):546-52, 1997.

Tunnicliffe WS, Burge PS, Ayres JG. Effect of domestic concentrations of nitrogen dioxide on airway responses to inhaled allergen in asthmatic patients. *Lancet,* 344(8939-8940):1733-6, 1994.

Venaille T, Snella MC, Holt PG, Rylander R. Cell recruitment into lung wall and airways of conventional and pathogen-free guinea pigs after inhalation of endotoxin. *Am Rev Respir Dis,* 139:1356-60, 1989.

Vogelzang PFJ, van der Gulden JW, Folgering H, Kolk JL, Heedrick D, Preller L. Tielen MJM, van Schayck CP. Endotoxin exposure as a major determinant of lung function decline in pig farm workers. *Am J Resp Crit Care Med,* 157:15-18, 1998.

von Mutius E, Martinez FD, Fritzsch C, Nicolai T, Roell G, Thiemann HH. Prevalence of asthma and atopy in two areas of West and East Germany. *Am J Resp Crit Care Med,* 149(2 Pt1):358-64, 1994.

Vukajlovich SW, Hoffman J, Morrison DC. Activation of human serum complement by bacterial lipopolysaccharides: structural requirements for antibody independent activation of the classical and alternative pathways. *Mol Immunol,* 24:319-31, 1987.

Wang JH, Devalia JL, Duddle JM, Hamilton SA, Davies RJ. Effect of six-hour exposure to nitrogen dioxide on early-phase nasal response to allergen challenge in patients with a history of seasonal allergic rhinitis. *J Allergy Clin Immunol,* 96(5 Pt 1):669-76, 1995.

Wang Z, Larsson K, Palmberg L, Malmberg P, Larsson P, Larsson L. Inhalation of swine dust induces cytokine release in the upper and lower airways. *Eur Resp J,* 10(2):381-7, 1997.

Weg VB, Walsh DT, Faccioli LH, Williams TJ, Feldman M, Nourhargh S. LPS-induced 111In-eosinophil accumulation in guinea pig skin: evidence for a role for TNF-alpha. *Immunology,* 84:36-40, 1995.

Weiss KB, Gergen PJ, Wagener DK. Breathing better or wheezing worse?

The changing epidemiology of asthma morbidity and mortality. *Ann Rev Pub Hlth,* 14:491-513, 1993.
Williamson IJ, Martin CJ, McGill G, Monie RD, Fennerty AG. Damp housing and asthma: a case-control study. *Thorax*, 52(3):229-34, 1997.
Yoshimura T, Matsushima K, Oppenheim JJ, Leonard EJ. Neutrophil chemotactic factor produced by lipopolysaccharide (LPS)-stimulated human blood mononuclear leukocytes: partial characterization and separation from interleukin-1 (IL-1). *J Immunol,* 139:788-93, 1987.

Chapter 10

THE ROLE OF GENETICS AND INDOOR ALLERGEN EXPOSURE IN ATOPIC ASTHMA

Kathleen C. Barnes, Ph.D. and Mark C. Liu, M.D.

I. INTRODUCTION

Persistent exposure to certain aeroallergens present in the indoor environment has been shown to be associated with chronic inflammation and asthma in genetically susceptible individuals who have been previously sensitized to the offending allergens. It can be argued that two of the important genetic components in atopic disease are the specific immune response genes, most notably the HLA-D region genes (which encode MHC molecules associated with the specificity of the response) and the overall genetic propensity of the individual to synthesize IgE. Although such genes are a necessary genetic requirement for atopic asthma, the manifestation of asthma is largely determined by environmental factors that up-regulate the expression of 'atopy' genes. This chapter focuses on what is known to date about the relevant genes contributing to atopic asthma and the most common indoor allergens involved in expression of atopic asthma.

II. ATOPIC ASTHMA

A. THE ROLE OF IgE

Asthma is one of several of the atopic diseases; however, there has been an ongoing debate about the precise contribution of IgE-mediated reactions to the causation of asthma. Is the non-specific bronchial hyperresponsiveness (BHR) observed in asthmatics caused by an intrinsic genetic 'defect,' or do allergens cause atopic reactions in the lungs ('extrinsic') which, in turn, lead to BHR? The observation that many non-specific factors may precipitate an asthma attack has been used to support the contention that atopy is not a central component in the causation of asthma. Alternatively, several physiologic phenomena related to allergen exposure typically occur in asthma (e.g., eosinophilia), and epidemiologic data suggest that the majority of asthma cases are associated with atopy.

In *cognate*, or specific IgE responsiveness, a major histocompatibility complex (MHC) class II molecule on an antigen-presenting cell (APC), such as a B cell, 'presents' a 'processed' antigen fragment to the T-cell receptor (TcR) on a Th2- or Th0-like cell (Marsh 1994; Marsh and Blumenthal 1990). This interaction leads to a specific IgE Ab response to the antigen that has previously been processed within the APC.

In contrast, there is a generalized polyclonal up-regulation of IgE production in the *noncognate*, or nonspecific response. Evidence suggests

1-56670-373-5/99/$0.00+$.50

that the interaction of basophils with B cells may be involved in this process, and other cells expressing FcεRI such as mast cells may also be involved (Bradding, et al. 1993; Gauchat, et al. 1993). The cytokine IL-4 is known to be important in immunoglobulin class-switching to IgE and in the differentiation of naïve T cells into Th2/Th0 cells (Bradding, et al. 1993; Gauchat, et al. 1993; Mossman and Coffman 1989; Romagnani 1992; Vercelli and Geha 1991). Th2/Th0 cells and basophils both produce IL-4, which provides a critical signal for IgE class-switching (via interaction with the IL-4R on neighboring B cells). Both cell types require IL-4 or IL-13 and can also express the CD40 ligand (CD40L), which interacts with CD40 on neighboring B cells to provide the necessary 'second signal' for IgE production (Gauchat, et al. 1993; Paul and Seder 1994; Vercelli and Geha 1991).

B. EPIDEMIOLOGIC STUDIES

More than 80% of patients with asthma have evidence of an allergic background (Braman, et al. 1987; Kapsali, et al. 1997; Smith 1983). Moreover, one can make the strongest case for an IgE-mediated causation of asthma in children, where virtually all asthmatics are skin test (ST) positive to implicated environmental allergens (Djukanovic, et al. 1990). Previously, a less persuasive case could be made in adults, especially older subjects, who are much less atopic than younger adults, and where 'intrinsic' asthma represents a substantial component of the clinically diagnosed disease (Djukanovic, et al. 1990). However, the data from the groups of Burrows et al. (Burrows, et al. 1989) and Sears et al. (Sears, et al. 1991), supported by others (Freidhoff and Marsh 1993; Tollerud, et al. 1991), provide a case for a major role of IgE (and, by implication, atopy) in adult asthma, including intrinsic asthma. In a large epidemiologic study in Tucson, Arizona, Burrows and associates (Burrows, et al. 1989) made the important observation that there is a clear relationship between the log of the risk for asthma and total serum IgE adjusted for age, sex, specific ST responsiveness to common inhalant allergens, and smoking habits. Alternatively, there was not as striking a correlation between the log of the risk for asthma and skin-test responsiveness to a panel of common environmental allergens. Burrows also showed that the presence of asthma is correlated with total IgE even in patients with negative skin tests; furthermore, no asthma was reported by the 177 subjects (7% of the study population) having the lowest range of IgE levels (≤ 2 IU/ml, or ≤ 5 ng/ml). Other groups have reported similar findings (Freidhoff and Marsh 1993; Sears, et al. 1991; Tollerud, et al. 1991).

III. INDOOR AEROALLERGENS

Persistent inflammation found in asthma is primarily (although not necessarily exclusively) the result of IgE responses to persistent exposure to small allergenic particles that are able to penetrate the lungs. Allergens associated with small particles in 'house dust' are likely to be universally responsible for inflammation in the lungs, because people spend most of their time living in indoor, often tightly sealed environments. Although outdoor, seasonal allergens such as pollen are clinically relevant, the worldwide increase in asthma prevalence and morbidity has been observed in people with perennial rather than seasonal disease (Platts-Mills and Carter 1997). Less is known about exposure to common allergens in public places; however, studies have found significant levels of dog, cat, dust mite, and cockroach allergens in the school setting (Einarsson, et al. 1995; Munir, et al. 1993; Sarpong, et al. 1997).

All of these allergens have been implicated in asthma, because they are known to be present continuously in the indoor environment in the form of small particles (usually <5 μM dia) that can readily be inhaled into the lungs. Persistent exposure to these allergenic particulates leads to localized wheal-and-flare, IgE-mediated reactions at the sites of deposition in the lungs about 15-20 minutes after allergen inhalation, often followed by more prolonged inflammatory late-phase reactions (LPRs) after about 4-6 hours. Although each allergenic particle is small, the highly localized concentration at the site of deposition is likely to be high—more than adequate to induce a local mini-LPR (It has been suggested that a small proportion of larger particles, such as whole mite feces (ca 20-25 μM dia) may enter the lung, perhaps under conditions of mouth-breathing, but earlier studies showing that whole ragweed pollen grains (18 μM dia) do not cause asthma in susceptible individuals (Bruce, et al. 1975) argue against this conjecture).

A. DISTRIBUTION AND IMPORTANCE

The two major groups of indoor allergens found in the domestic environment are arthropod-borne allergens (e.g., dust mites, cockroach) and allergens from the epithelia/saliva/urine of household pets, notably cats (Colloff, et al. 1992; Platts-Mills, et al. 1992). Additional indoor allergens include rodents, and, to a lesser extent, mold spores and mycelia (Burge 1989; Desjardins, et al. 1993; Hamilton, et al. 1992; Lehrer, et al. 1994). Many of the major and minor allergens from mites, cat, and cockroach have been purified, characterized and studied intensively (eg, Chapman, et al. 1988; Chapman, et al. 1987; Heymann, et al. 1989; Platts-Mills, et al. 1997; Pollart, et al. 1991a). In the case of mites (*Dermatophagoides* spp., *Blomia tropicalis*), cockroach (*Blattella germanica*, *Periplaneta americana*), cat and dog, a number of the allergens have been cloned (Platts-Mills, et al. 1997). The availability of these allergens has permitted the development of highly sensitive enzyme-linked immunoassays (ELISAs), often using monoclonal antibodies, for the quantitative determination of the relevant major allergen(s). Such assays, pioneered by Platts-Mills, Chapman and associates (Platts-Mills, et al. 1985) are highly reproducible and have proved to be extremely valuable in simplifying the measurement of mite, animal dander and cockroach allergen levels in the indoor environment.

B. MITES

Mites of the genus *Dermatophagoides* (*e.g.*, *D. pteronyssinus*, *D. farinae*) are considered to be the primary indoor allergens worldwide; however, in mite-free environments, allergens such as cat may be the most important (Sporik, et al. 1995) and sensitization to them is highly associated with asthma (Ingram, et al. 1995). Mites are microscopic arachnids which feed primarily on human dander and prefer warm, moist environments typically found in the bed and in the padded furnishings of the house. Under favorable growth conditions, house dust mites produce many times their bodyweight of allergenic particles in the form of feces and body parts, which persist in the dust, where they may be broken down into smaller particles. A thorough review of house dust mites and their importance in allergic disease has been put forth as a result of three international workshops (Platts-Mills, et al. 1997; Platts-Mills, et al. 1992; Platts-Mills and Weck 1989).

Blomia tropicalis is one of the storage mite species common in the subtropical and tropical regions of the world, where the *B. tropicalis* allergen has been found in close to 100% of house dust samples (Fernandez-Caldas, et al. 1993b; Neffen, et al. 1996; Puerta, et al. 1996; Tsai, et al. 1998). Similarly, sensitization to *B. tropicalis* has been detected in numerous locales, yielding some of the highest number of positive responses by skin test (ST) and specific IgE compared to other mites (Fernandez-Caldas, et al. 1993a; Ferrandiz, et al. 1996; Neffen, et al. 1996). It is likely that, in the tropics, the allergenic role of *B. tropicalis* may be as important as that of the *Dermatophagoides* spp.. The major allergens of *B. tropicalis* have been identified, the most important of which is Blo t 5, which shares 43% sequence homology to Der p 5, but which has unique allergens capable of eliciting specific immunoresponsiveness (Arruda, et al. 1997).

C. COCKROACHES

Numerous studies have been published which confirm the relationship between cockroach allergens and bronchial asthma (Bernton and Brown 1970; Gelber, et al. 1993; Kang 1976; Kang, et al. 1992; Pollart, et al. 1991b; Rosenstreich, et al. 1997). The prevalence of asthma directly related to cockroach allergens among atopic asthmatic children has been reported to be as high as 69% (Kang and Chang 1985). More importantly, asthmatics sensitive to cockroach allergen are more severe asthmatics than the general asthma population, exhibiting higher IgE levels, a higher degree of steroid dependency and greater chronicity of disease (Kang, et al. 1993; Kang, et al. 1992).

Cockroach infestations are typically highest in households of the lower socioeconomic strata (Brenner, et al. 1987; Woods, et al. 1981), and, not surprisingly, cockroach allergy is typically of greatest significance among asthmatics who live in poor (especially inner city) housing (Gelber, et al. 1993; Horowitz, et al. 1996; Pollart, et al. 1989; Rosenstreich, et al. 1997). The primary species of interest are the domestic *Blattella germanica* ('German' cockroach) and *Blatta orientalis* ('Oriental' cockroach) and the peridomestic *Periplaneta americana* ('American' cockroach). Four of the major allergens for *B. germanica* (Bla g 1, Bla g 2, Bla g 4, Bla g 5) and one of the major allergens for *P. americana* (Per a 1) have been identified (Arruda, et al. 1995; Chapman, et al. 1997; Helm, et al. 1994; Pollart, et al. 1991a; Schou, et al. 1990). Because of its omnivorous nature, the cockroach can survive on virtually any product consumed, discarded, or excreted by humans. The low-income environment presents an ideal habitat for a pest which thrives on moisture (e.g., poor disposal of garbage, leaky plumbing), low air flow (e.g., abundant cracks and crevices and dead spaces between dwellings), and warmth (e.g., poorly ventilated units). Allergen concentrations are typically highest in kitchens, but may be present in significant concentrations in the bedroom (Gelber, et al. 1993; Mollet, et al. 1997; Pollart, et al. 1991b). Brenner et al. (Brenner, et al. 1987) reported that, in a low-income housing project in Florida, the mean number of German (*Blattella germanica*) cockroaches was 13,163.

D. PETS

It is estimated that approximately 38% of American households own a dog *(Canis domesticus)* and 31% own a cat (*Felis domesticus*) (Troutman 1988a; Troutman 1988b). A number of allergens are produced by the domestic cat, but the most clinically relevant in terms of exposure is Fel d 1, a 36 kDa protein first characterized by Ohman and others (Ohman and Lowell 1977). Fel d 1 is produced primarily in the intradermal hair follicles and collects on the fur (Charpin, et al. 1991; Mata, et al. 1992) and is also produced by the salivary glands (Anderson and Baer 1981; Leitermann and Ohman 1984). It is unique among the common indoor allergens in that it is both a'sticky' allergen, readily collecting on walls, upholstered furniture, carpets, and even dust particles, and it is a stable antigen, maintaining potency for up to 20 weeks or greater after removal of the pet from the home (Wood, et al. 1989). In homes where cats are present, levels of Fel d 1 allergen have been found to be as high as 10 μg/gram of dust (Chapman, et al. 1988) and recently it was shown that, even in homes without a cat, levels of allergen which are capable of inducing upper and lower respiratory symptoms are present (Bollinger, et al. 1996).

Domestic dog allergen is, similarly to domestic cat allergen, ubiquitous in households owning a dog, yet the spectrum of important allergens is more complex than that for the cat. For example, up to 28 distinct allergens were initially identified by Ford and collaborators (Ford, et al. 1989). The major allergen, Can f 1 (25-kDa) was subsequently purified and characterized. Can f 1 is produced in the hair, dander, and also in the saliva (Ford and Kemeny 1992; Konieczny, et al. 1997; Schou, et al. 1991b). Schou and others (Schou, et al. 1991a) collected dust samples from 103 U.S. homes and found that, in homes where a dog was present, the median level of Can f 1 allergen was 120 μg/gram of dust, compared to a median of 3 μg/gram of dust in homes without a dog.

E. DOSE-RESPONSE STUDIES

It is fairly well-established that a dose-response relationship exists between exposure to certain levels of the most common indoor aeroallergens and sensitization (*see Refs*. Hamilton, et al. 1992 and Platts-Mills, et al. 1997 *for review*), especially for Der p 1 (Call, et al. 1992; Peat, et al. 1994; Rosenstreich, et al. 1997; Sporik, et al. 1995; Squillace, et al. 1997; Wahn, et al. 1997; Wickman, et al. 1991). The association between sensitization to certain allergens and *asthma* is less conclusive, although quite often mite sensitization is the most important risk factor for asthma (Burrows, et al. 1995; Nelson, et al. 1996; Peat, et al. 1996; Squillace, et al. 1997). It also appears that the most important allergen associated with asthma varies according to whatever happens to be the dominant allergen in that particular locale. For example, in Los Alamos, New Mexico, where the climate is incompatible with house dust mite survival, but where high levels of cat and dog allergen were found, it was shown that sensitization among asthmatic children to those allergens was very strongly associated with asthma symptoms and BHR (Platts-Mills, et al. 1995a). Among asthmatics living in U.S. inner cities, sensitization to allergens from the German cockroach is an important risk factor (Call, et al. 1992; Gelber, et al. 1993; Rosenstreich, et al. 1997; Sarpong, et al. 1997) and may even be associated with asthma severity (Horowitz, et al. 1996; Rosenstreich, et al. 1997).

Threshold levels for several of the indoor allergens have been proposed for the purpose of assessing a cause-and-effect relationship between exposure and asthma, although recently it has become evident that this particular dose-response relationship is rather complex (Platts-Mills, et al. 1995b). For example, a threshold level for Der p 1 has been proposed at 2 μg/gram of dust (Platts-Mills, et al. 1992; Platts-Mills and Weck 1989) which roughly corresponds to the $\geq$100 mites/gram of dust threshold proposed by Voorhorst and collaborators in 1969 (Voorhorst, et al. 1969). And in fact, a number of studies have, to date, demonstrated a relationship between chronic exposure to Der p 1 levels in excess of 2 μg/gram of dust and various parameters associated with asthma, such as symptoms and morbidity (Chan-Yeung, et al. 1995; Kivity, et al. 1993; Squillace, et al.

1997; Vervloet, et al. 1991). For example, a strong, significant correlation ($r = 0.66$; $p = 0.001$) was established by Sporik et al. (Sporik, et al. 1990) between the quantitative level of Der p 1 in house dust samples in the U.K. and the age of onset of asthma in children from 'asthmatic' families. These findings and others suggested a causal relationship between the presence of mite allergens and the *development* of asthma. Such clinically-relevant, threshold levels of the major indoor allergens also include Der f 1 (*D. farinae*) ≥ 2 μg/gram of dust, Fel d 1 ≥ 8 μg/gram of dust, and Bla g 1, Bla g 2, Per a 1 >1 Units/gram of dust (*see* Hamilton, et al. 1992 *for review*).

Other studies suggest that the dose-response relationship is not so straightforward. Munir and collaborators (Munir, et al. 1997) showed that Swedish children exposed to *below*-threshold levels of indoor allergens but who had a clear family history of atopy were at risk for sensitization to the relevant allergen. Alternatively, Marks and others (Marks, et al. 1995) found that, even though nearly all of the schoolchildren whom they studied in Sydney were exposed to the highest concentrations of Der p 1 allergen ever reported (mean=51.0 μg/g dust), there was no association between allergen concentrations and asthma symptoms, BHR, or skin test reactivity. In a large U.S. inner-city study of 476 children with asthma it was found that a dose-response relationship existed between exposure to high levels of cockroach allergen and asthma morbidity, but among those same children there was no relationship between exposure to high levels of mite or cat allergen and asthma morbidity, despite positive skin tests to the allergens (Rosenstreich, et al. 1997). Platts-Mills and others (Platts-Mills, et al. 1995b) have postulated that several factors may account for the sometimes poor relationship between allergen exposure and asthma symptoms and/or morbidity, including (1) the accuracy to which dust samples collected in a conventional manner reflect exposure; (2) the relevance of specific IgE measurements (e.g., skin test, serum IgE antibody) as a measure of inflammatory lung disease; and (3) the accuracy to which disease severity reflects pulmonary inflammation. Furthermore, they suggest that the dose-response relationship is not necessarily linear, but perhaps it is log linear, or sigmoidal, and that potentially much higher (or much lower) levels of allergen than the proposed threshold are required to affect morbidity (Platts-Mills, et al. 1995b).

IV. GENETIC STUDIES OF ATOPY AND ASTHMA

As discussed above, high levels of mite allergens (even above the level of 2 μg Der p 1/gm of dust) fail to sensitize most people, and some people who are highly sensitized develop allergic rhinitis, but no asthma. Why is this? It can be argued that only *genetically susceptible* individuals, with the 'right' combination of atopy and, possibly, asthma-specific genes, become asthmatic and only when adequately and persistently exposed, usually to the cited indoor allergens of which the *Dermatophagoides* mites are of special interest.

The allergens responsible for the inflammation seen in the lungs of asthmatic subjects may or may not be the same as the allergens that precipitate an asthmatic attack, since asthmatics often have multiple atopic sensitivities. The old experimental model of asthma using bronchial provocation with allergenic extracts, which is really 'a large skin test performed in the lung' with a non-physiologic, extremely high allergen dose, provides evidence for atopic sensitization that can more easily and safely be obtained by skin testing. With respect to atopy, the real epidemiologic-genetic questions are, "Which allergens primarily cause the underlying inflammatory response and why does this occur only in certain individuals?"

In view of the considerations raised above, it is not surprising that genetic studies of the allergic diseases have met with limited success (Marsh and Blumenthal 1990; Marsh, et al. 1993). Borecki et al. (Borecki 1985) attempted to relate the genetics of IgE regulation to the different manifestations of allergic disease using total IgE and clinical data from the 173 families studied earlier by Gerrard et al. (Gerrard, et al. 1978). They showed that a major locus determining high IgE levels markedly influences the development of different manifestations of atopic disease. Specifically, when only asthma and atopic dermatitis were considered, the effect of the major gene was compatible with recessive inheritance. Cookson and Hopkin (Cookson and Hopkin 1988) proposed autosomal dominant inheritance of atopy, and in a subsequent genetic linkage analysis (Cookson, et al. 1989) provided evidence that their 'atopy gene' is linked to the *D11S97* marker in chromosome 11q13.

A. HYPOTHESES

Although findings from numerous groups have demonstrated that there is a strong heritable component to asthma and atopy, the genetic pathophysiology of these traits is poorly understood. Studies in which the concordance for atopic disease was greater between monozygotic (MZ) twins compared to dizygotic twins provided some of the first evidence that allergic disease had a familial basis (Cookson, et al. 1989; Lympany, et al. 1992; Marsh, et al. 1993). However, a concordance of only 50-60% for atopy (Marsh, et al. 1981) suggest that, in addition to the role of environmental influences in the expression of disease, which could account for some of the disparity of disease expression, gene mutations and/or rearrangements are contributing to the complex nature of allergic disease. This is particularly pertinent in view of the findings that MZ twins raised in presumably the same environmental setting frequently exhibit similar total serum IgE levels, but disparate *specific* IgE Ab responsiveness (Marsh, et al. 1981).

From these studies as well as the work by Burrows and colleagues (Burrows, et al. 1989), there appear to be at least three main categories of genes controlling the expression of atopic disease: (1) cognate, specific immune response genes, which determine the ability of the individual to make specific immune responses, dictated in part by environmental factors (e.g., the level of the person's exposure to indoor allergens); (2) noncognate, or nonspecific atopic genes that determine the overall propensity of the individual to synthesize IgE of any specificity; and (3) airway inflammation genes, which may be separable from those determining atopic disease overall.

B. GENETICS OF COMPLEX TRAITS: PRINCIPLES AND PRACTICES

Atopy has been described as a complex trait for which genetic predisposition is fundamental in disease etiology, but poorly understood, largely due to the myriad of potential phenotypes (e.g., severity, IgE levels, pulmonary function). In this sense, atopy is *genetically heterogeneous*, because many distinct genes, or groups of genes, are undoubtedly involved in the expression of disease, resulting in different types of disease manifestation, or the differential expression of certain characteristics of the overall phenotype.

Unraveling the genetic etiology is further complicated by the tremendous influence of the environment over disease expression. For this reason, atopy is *multifactorial*, in that disease expression is influenced by interactions between multiple major and minor genes, and it is modulated by interacting non-genetic factors, such as the environment. The multifactorial nature of asthma and atopy can best be conceptualized by imagining multiple interactions of genes, the environment, and, to a lesser (and poorly understood) degree, developmental "noise" (e.g., the fetal environment, respiratory infections early in life, immunoglobulin-T cell receptor gene rearrangements).

It is critical to consider, given that an individual has the *genetic propensity* to develop asthma, that she or he must subsequently be exposed to certain environmental factors, such as allergens, in order to express the disease. In short, genes and the environment are *necessary* for the expression of disease, but by themselves are *not sufficient* for the expression of disease. This factor complicates the search for asthma genes, because one really cannot be sure that a person has the gene for asthma if that individual has never had sufficient exposure to the allergens necessary to elicit the response.

The Candidate Gene Approach. Genetic mapping, the method by which the inheritance pattern of a trait is compared with the inheritance pattern of chromosomal regions, was revolutionized with the availability of highly polymorphic, genetic markers in the 1980's (*see ref.* Lander and Schork 1994 *for review*). These 'flags' or 'sign-posts', which indicate the approximate location of the genes suspected of contributing to a particular disease, have aided in the isolation of polymorphisms, or variants within a given allele, that play a role in different expressions of a particular phenotype, or disease.

The primary method used to identify the genes responsible for asthma and atopy is the *candidate gene approach*, in which the gene causing a human disease is identified based on partial functional information (e.g., the role of IL4 in atopic disease), in order to make an educated guess as to which gene or genes one should be focusing. From there the process of marker analysis is begun in order to visualize the polymorphisms involved, and, using segregation analyses, to look for a relationship between the gene of interest and expression of the disease.

An early example of the candidate gene approach with a focus on atopy is the study conducted by Marsh and his colleagues which relates to evidence that a gene(s) in human chromosome 5q31–q33 exerts control of IgE production and may play a role in the expression of asthma (Marsh, et al. 1994). Chromosome 5q31–33 contains multiple candidate genes for allergy and asthma (Marsh, et al. 1994; Saltman, et al. 1993), including the clustered family of cytokine genes (*IL3, IL4, IL5, IL13* and *CSF2* [*GMCSF*]) which are well-known for their interactive roles in the allergic inflammatory response (Paul and Seder 1994). After selecting a region for which several attractive candidates for the genetic regulation of serum IgE were located, the group analyzed *IL4-R1* (located within the second intron of the *IL4* gene) and seven other polymorphic DNA markers in this region using genomic DNA from 11 extended Amish pedigrees, comprising 20 nuclear families (170 subjects and 349 sib pairs) selected for at least one child with IgE antibodies to one or more common inhalant allergens. Using sib-pair linkage analysis (Lander and Schork 1994; described below) they found evidence for linkage with total IgE, but not specific IgE, within the 5q31.1 region. Marsh's study and studies by other groups (Doull, et al. 1996; Meyers, et al. 1994; Postma, et al. 1995; Walley, et al. 1997) suggested that *IL4* and/or nearby gene(s) in 5q31.1 regulates IgE production in a noncognate fashion. Subsequently, a polymorphism (T variant at position –590) in the *IL4* promoter was found to be associated with elevated total serum IgE levels in families with asthma (Rosenwasser, et al. 1995).

The evidence for linkage on 5q is some of the most solid to date with regard to the genetics of atopy, but clearly work is necessary to define the functional relationships in more detail and certainly we cannot rule out the involvement of other genes in the region.

Genome-Wide Search Approach. With the development of positional cloning (Bender, et al. 1983), a process by which linkage is established between the presence of the disease/disease-associated phenotype(s) and highly polymorphic genetic markers, geneticists could isolate a gene of interest without knowledge of its biochemical function. This alternative approach, sometimes called a *genome-wide search*, involves the scanning of the entire genome with a dense collection of genetic markers (~300-400) typically spaced 10-20 cM apart (1 cM = about 1 million base pairs). As with the candidate gene approach, the appropriate linkage statistic S(x) is calculated at each position x along the genome, and the regions in which the statistic S shows a significant deviation from what would be expected by chance are identified.

The first results of a genome-wide search for susceptibility loci linked to asthma was reported in 1996, which showed evidence for linkage to asthma and atopy in five novel regions in families from Australia (Daniels, et al. 1996). Shortly thereafter, the U.S.-based, NHLBI-funded *Collaborative Study on the Genetics of Asthma (CSGA)* published their findings from a genome-wide search using families from three ethnic groups (Caucasian, African American, Hispanic) and reported linkage to six other novel regions plus five previously identified regions (Anonymous 1997). Interestingly, the combined genome-wide scans in the Cookson study and the *CSGA* found linkage in only one common region (13q). Of greater concern, Cookson's group did not find evidence of linkage for two of the five new loci in a second panel of families from the U.K., nor did they find evidence of linkage to several previously-identified regions (e.g., 5q). Similarly, the *CSGA* genome-wide search did not provide evidence of linkage in more than one ethnic group for any of the six novel regions.

Analytical Strategies. Several different nonparametric approaches (approaches which assume no particular pattern of inheritance, e.g., dominant, recessive) have been used to dissect the complex allergic diseases. The 'allele-sharing method' is an approach whereby one estimates the proportion of marker alleles shared identical by descent (IBD) between family members (e.g., siblings, relatives) for a given phenotype as a test for linkage to unobserved susceptibility loci (Lander and Schork 1994; Risch 1990). This method is widely used for linkage analysis of complex phenotypes, both qualitative (disease vs. no disease) and quantitative (continuous traits, such as IgE). Essentially, this approach tests the null hypothesis of no linkage by comparing the proportion of marker alleles shared IBD within each pair of relatives (usually siblings) to that expected for a marker completely unrelated to the phenotype (e.g., the Mendelian expectation of sharing 0, 1, or 2 alleles IBD for any pair of sibs is 0.25, 0.50, and 0.25, respectively, under the null hypothesis). The alternative hypothesis of linkage predicts that concordant sibs should share more marker alleles IBD than discordant sibs, while linkage for a quantitative phenotype predicts sibs with more similar phenotypic values should share more marker alleles IBD (as tested in a linear regression model). The methodological approach used most often to perform two-point, genetic linkage analysis is *sib-pair analysis* (Haseman and Elston 1972). Increasingly, however, multipoint affected relative analyses (based on multipoint IBD) are used and they have the advantage of summarizing evidence for linkage over entire chromosomal regions, using both log of the odds (LOD) score statistics under specified models of inheritance and a nonparametric approach.

Association studies are yet another approach in understanding the genetics of atopy, but, unlike allele-sharing methods, association studies are case-control studies, comparing affected and unaffected, *unrelated* individuals. Rather than focusing on inheritance patterns, association studies test whether a specific allele at a gene (or marker) occurs at a higher frequency among individuals affected with the trait of interest compared to unaffected individuals (Lander and Schork 1994).

Both linkage disequilibrium and association studies can be performed using the *transmission disequilibrium test*, or TDT (Ewens and Spielman 1995). The TDT has the advantage of combining a linkage approach and a population approach (e.g., not requiring multiplex sibships and/or additional family members), and tests for excess transmission of marker alleles to affected children. The basic design of the TDT is to compare marker alleles transmitted to an affected child from a heterozygous parent to those not transmitted. To perform the test the number of transmissions and non-transmissions of alleles from heterozygous parents to their affected offspring are compared. The number of times a particular allele is transmitted to offspring is compared to an expected ratio of 1:1, and evidence of linkage disequilibrium is determined if, in a given population of these trios of mother, father and child, the number of times an allele is transferred is significantly greater than or less than that which would be expected by chance.

C. GENETIC CONTROL OF THE SPECIFIC IMMUNE RESPONSE

The genes involved in specific immunoresponsiveness can be divided into two types: (1) genes that are directly involved in antigen recognition (e.g., *HLAD*, *TCR)*, and (2) genes that are part of the cognate response *independent* of the precise antigenic specificity (e.g., TCR/CD3 complex signal-transduction, *IL4* gene-regulation).

Antigen Recognition Candidate Genes. Class II molecules important in antigen presentation and the human leukocyte antigen (HLA) complex are encoded by genes in the HLA-D region of the human major histocompatibility complex (MHC) located on the short arm of chromosome 6 (6p21.3). Some of the earliest studies on the genetics of atopy focused on the HLA-D subregions (DR, DQ, DP) of the human MHC and were conducted by David Marsh and his group to examine immune responsiveness to two of the major allergens of short ragweed, Amb a 5 (5 kDa) and Amb a 6 (9.9 kDa) (Huang, et al. 1991; Marsh, et al. 1990; Zwollo 1991). Marsh postulated that immune recognition involved the interaction of a single MHC class II molecule (or a group of closely related class II molecules) with a single major T-cell epitope of the antigen. Based on this assumption, it was expected that unrelated individuals would share a role for *HLAD* genes in cognate responses. In fact, the group found a class II restriction limited to individuals carrying HLA-DR2 and Dw2 (DR2.2) in all but two of approximately 80 Caucasian IgE responders to Amb a 5. However, in atopic individuals who did not respond to Amb a 5, the DR2.2 frequency was similar to the Caucasian population at large (~22%).

The group went on to identify the actual class II restriction elements associated with the DR2.2 and DR2.12 phenotypes in Caucasians (Dr$\alpha\beta$1*1501 molecule) and Asians (DR$\alpha\beta$1*1502 molecule), respectively. These studies clearly showed a cause-and-effect relationship between the presence of a particular HLA-D phenotype with specific IgE and IgG Ab responsiveness. Subsequently, numerous studies confirmed Marsh's findings, and HLA-D associations have been extended to other allergens important in asthma, including giant ragweed, perennial rye, and also the mite allergens Der p 1 and Der p 2 (Blumenthal, et al. 1992; Coulter, et al. 1987; Freidhoff, et al. 1988; Marsh 1994; O'Hehir, et al. 1991; Okano, et al. 1996; Roebber, et al. 1985). Collectively, these results provided a clear demonstration of the role of HLA-D-encoded molecules in well-defined immune responses toward simple allergen molecules which appear to possess single major epitopes.

Linkage studies in this region have been less conclusive, are very few, and, not surprisingly, have focused primarily on mite allergy. In a study of 20 families from Cartagena, Colombia, with mite allergic asthma, Caraballo and collaborators (Caraballo and Hernandez 1990) provided data suggesting the existence of HLA-linked genes controlling mite-specific IgE concentrations. Subsequently they determined that HLA-DR3 is associated with *B. tropicalis*-specific IgE immune responsiveness (Caraballo, et al. 1996). As part of the CSGA genome-wide screen, evidence for linkage of Der p 1-specific IgE responsiveness and chromosome 6p21 markers in the *HLAD* region was found (P=0.0064) among a group of 45 Caucasian families (Hizawa, et al. 1998b). Using immunoblotting analysis to evaluate specific IgE immune responses toward the Der p allergens, the group subsequently identified two alleles (the 196 base pair [bp] of *D6S1281*, the 104 bp of *DQCAR*) which showed excessive transmission to Der p-specific IgE responders toward a particular Der p polypeptide (Hizawa, et al. 1998a).

With regard to the expression of specific IgE responsiveness to the common indoor allergens (and, for that matter, outdoor allergens), the HLA region is considered to be a necessary, but not sufficient factor (Blumenthal and Amos 1987; Zwollo 1991). Other non-*HLA* genes have been shown to contribute to the expression of the cognate response. A very early family study of IgE concentrations and quantitative skin tests illustrated that genes regulating total IgE concentrations affected specific IgE-mediated sensitivity and were capable of masking the effect of the *HLAD* gene (Marsh, et al. 1974). In another study, it was found that strong association between HLA-DR3 and Lol p 3- (a major allergen of the grass pollen *Lolium perenne*) specific immune responsiveness was less striking in responders who had high total IgE concentrations compared to responders who had low total IgE concentrations (Ansari, et al. 1989). More recently, Hizawa and others (Hizawa, et al. 1998c) showed evidence for linkage to skin test reactivity to dog and markers in 5q31—q33 (P=0.0043), and skin test reactivity to *D. farinae*, cat, and Bermuda grass and markers in 11q13 (P=0.012, P=0.035, P=0.017, respectively) among African American subjects participating in the *CSGA*.

Within a distance of 5 cM from *HLAD* is the gene encoding tumor necrosis factor (*TNF*). The close proximity of these two genes suggests that it might be difficult to tease out, using marker analysis, precisely which gene is in linkage disequilibrium with the specific immunoresponsiveness phenotype. Nevertheless, in a study of 88 unrelated families, Moffatt and the Oxford group (Moffatt and Cookson 1997b) were able to distinguish the genetic effects of TNF from those of HLA-DR, and in fact showed an association with asthma and two TNF polymorphisms (Ltα*Nco*I*1, TNF-308*2), but not with HLA-D.

Also clustered in the HLA class II region is the gene encoding transporter antigen peptide 1 (*TAP1*), an interferon-γ—inducible gene involved in antigen processing (Spies, et al. 1990). In their studies of a group of atopic and normal Tunisian subjects, Ismaïl and others (Ismaïl, et al. 1997) recently investigated the potential association of a *TAP1* polymorphism with atopy. In their study they demonstrated an increased relative risk of atopy among individuals who carried a glycine at position 637 of the TAP1-B and TAP1-D alleles, and there was an especially high relative risk ($P \leq$ 0.0001) of atopic asthma for homozygotes (TAP1 D/D; Ismaïl, et al. 1997). The D allele, however, was not observed in subjects of European descent. It will be crucial to replicate these findings in other populations in order to better understand the role of *TAP1* in the pathogenesis of atopy.

T-Cell Receptor alpha/delta Complex. In addition to the HLA, a molecule that is essential in foreign antigen recognition and handling is the T cell receptor (TcR). The majority of T cell receptors are comprised of alpha and beta chains, and the genes encoding these products have been identified and are located on chromosomes 14 and 7, respectively. To a much lesser extent these receptors are made up of gamma and delta chains (the delta chain genes are located within the alpha chain locus). Given the tremendous potential for variation that arises from the many variable (V) segments within the TcR and the important role that the peripheral TcR repertoire plays in the genetic control of immune responsiveness (Loveridge, et al. 1991), this is yet another intriguing candidate for the genetic study of the cognate response.

Using sib-pair analysis on two sets of families from the U.K. and Australia, Moffatt, Cookson and others (Moffatt, et al. 1994) showed evidence for linkage of specific IgE responsiveness to the TcR-α/δ locus on chromosome 14q11. Because it had previously been shown that the V region gene segment Vα8 was dominant in the TcR recognition of house dust mite-specific CD4+ T-cell clones (Wedderburn, et al. 1993), the Oxford group went on to perform association studies of the Vα8.1 polymorphism and IgE concentrations to Der p (Moffatt, et al. 1997). In a study of two sets of subjects from their previous linkage study, they found a strong allelic association between allele 2 of the Vα8.1 polymorphism (Vα8.1*2) and reactivity to Der p 2, but only a weak association with Der p 1. In further analyses they found an enhanced response to Der p 2 in subjects who were positive to both Vα8.1*2 and HLA-DRB1*1501, indicating that the interaction of these two elements modifies the response to exogenous antigen (Moffatt and Cookson 1997a; Moffatt, et al. 1997).

D. GENETIC CONTROL OF THE NONSPECIFIC IMMUNE RESPONSE

IL-4 Gene Cluster. The evidence for linkage of markers in the IL-4 gene cluster and total IgE production was described above (IV.B.) as part of an illustration of the *Candidate Gene Approach*. Briefly, because the cytokine IL-4 is central in both cognate and noncognate processes of IgE-switching, as well as its known role in Th2-cell development, the *IL4* gene, located in chromosome 5q31–q33, is an appropriate candidate for studying the genetics of IgE production. Also as described above, there are other intriguing candidates clustered in this region, including *IL3, IL4, IL5, IL13* and *CSF2* (*GM-CSF*). Nearby (telomeric) are the glucocorticoid receptor (*GRL*) and β2-adrenergic receptor (*ADRB2*) genes. As it stands currently 5q31–q33 is one of the regions for which there is the strongest evidence of linkage to asthma and atopy. Linkage of total IgE concentrations to the 5q31–q33 region is supported by the studies of Marsh (Marsh, et al. 1994) and reports from the laboratories of Meyers et al (Meyers, et al. 1994; Postma, et al. 1995), Rosenwasser and Borish (Rosenwasser, et al. 1995), Cookson (Walley, et al. 1997) and Holgate (Doull, et al. 1996) are consistent with linkage more specifically to 5q31.1 and *IL4* region and associated phenotypes. More recently, evidence of linkage to the clinical phenotype ‘asthma’ was found in Caucasians participating in the CSGA genome-wide search project and 5q23–q31 (Anonymous 1997) (Mlod=0.94, P=0.0187).

The results from these studies suggest that *IL4*, or a nearby gene in 5q31.1, is involved in the regulation of IgE production in a nonantigenselective – or noncognate fashion. Basophils, in particular, may be implicated in this non-cognate IgE production, although other FcεRI-positive cells may also be involved. It seems likely that a generalized up-regulation of IL-4 could induce B cells (that are precommitted to make IgG antibody to a broad array of antigens) to switch to IgE. Under normal physiology, such 'polyclonal' IgE would not be specific for common environmental allergens.

β Subunit of the IgE High-Affinity Receptor (FCERIB). Another region which contains a likely candidate associated with asthma and airway inflammation and one for which there is compelling evidence for linkage is 11q13. The first evidence for an 'atopy' gene linked to chromosome 11q13 was by Cookson's group at Oxford (Cookson, et al. 1989). Following a preliminary screen of possible linkage to 17 polymorphic markers scattered throughout the genome, the group reported evidence for linkage of atopy to 11q13, and subsequently provided further evidence for linkage in two separate populations (U.K. and Australia) as part of their genome-wide search (Daniels, et al. 1996). The same group subsequently showed that an amino acid substitution in the FcεRI-β chain on chromosome 11q13 segregated with 'atopy' in a small group of families from the U.K. (Cookson, et al. 1992; Shirakawa, et al. 1994b). To date a number of groups have confirmed this linkage (Boguniewicz and Hayward 1996; Collée, et al. 1993; Hizawa, et al. 1995; Shirakawa, et al. 1994a; Young, et al. 1992). Cookson and his colleagues have also identified a variant in exon 7 (E237G) which renders a relative risk of asthma of 2.3 in carriers (P=0.005) compared to subjects who do not have the polymorphism (Hill and Cookson 1996). It has been proposed that the *FCER1B* gene is the primary candidate in this region (Daniels, et al. 1996; Shirakawa, et al. 1994b). *FCER1B* encodes the β chain of the high-affinity receptor for IgE and therefore implicates noncognate IgE production.

Interferon-gamma and other candidates in 12q. Relatively recent studies indicate linkage to asthma and associated phenotypes on 12q15–q24.1 (Barnes, et al. 1996; Campbell, et al. 1997; Nickel, et al. 1997; Wilkinson, et al. 1997; Wilkinson, et al. 1996). This region contains several candidate genes potentially important for asthma susceptibility; of particular interest is the gene encoding interferon-γ (*IFNG*), the cytokine which opposes the effects of IL-4. Interestingly, it was recently reported that a CA repeat polymorphism in the first intron of the *IFNG* gene correlates with a higher production of IFN-γ (Asderakis, et al. 1997). Other candidates include stem cell factor (*SCF*; previously referred to as mast cell growth factor, or *MGF*), insulin-like growth factor-1 (*IGF1*), the β subunit of nuclear factor-Y (*NFYB*), leukotriene A4 hydrolase (*LTA4H*), and B-cell translocation gene 1 [*BTG1*] (Krauter, et al. 1995). Also of interest in this region is *STAT6*, the gene important in the IL-4/IL-13-mediated signal transduction pathway.

The first evidence for linkage to asthma and the associated phenotype 'high total IgE' to markers in this region came from sib-pair analyses of two diverse populations (Barnes, et al. 1996), the isolated, Caucasian Amish and multiplex Afro-Caribbean families from the West Indian island of Barbados. Similar findings were obtained using multipoint IBD analysis. These findings have since been supported by other published studies, including the *CSGA* (Anonymous 1997), the Southampton group (Wilkinson, et al. 1997) and a well-designed study in an outbred German group of children and their parents (Nickel, et al. 1997). In the German study, 52 children with persistently 'high' total serum IgE (defined as $\geq$ 85th percentile of the tested population, documented at $\geq$ 2 time points) and their parents were genotyped with nine polymorphic markers, and data were analyzed using the TDT (described in IV.B.). Evidence for linkage and allelic association for high total IgE was observed for three of the markers for which evidence of linkage had been detected in the Barbados dataset. The findings from the German study demonstrated the value of using extreme phenotypes in genetic analysis of a complex quantitative trait (IgE), and further substantiated the initial linkage to 12q.

V. CONCLUSION

The collective findings from the various studies described in this chapter underscore the profound complexity of the genetics of asthma, atopy, and associated traits, and suggests that the effects of ethnicity as well as the influence of environmental factors must be considered when conducting this work. It is important to note that none of the genetics of asthma and atopy studies (candidate-gene or genome-wide search) have focused on the role of environmental exposure as a covariate in the dissection of the disease. Given the voluminous information available to us at this time regarding the critical role of environmental exposure and expression of disease, new approaches in the genetic dissection of asthma and atopy are certainly warranted.

REFERENCES

Anderson, M.C. , and H. Baer. 1981. Allergenically active components of cat allergen extracts. *J Immunol* 127(3):972-975.

Anonymous. 1997. The Collaborative Study on the Genetics of Asthma: A genome-wide search for asthma susceptibility loci in ethnically diverse populations. *Nature Genetics* 15(4):389-392.

Ansari, A.A. , et al. 1989. Human immune responsiveness to Lolium perenne pollen allergen Lol p III (rye III) is associated with HLA-DR3 and DR5 [published erratum appears in Hum Immunol 1989 Oct;26(2):149]. *Hum Immunol* 25(1):59-71.

Arruda, L.K. , et al. 1995. Molecular cloning of German cockroach (Blattella germanica) allergens. *Intl Arch Allergy Immunol* 107(1-3):295-297.

Arruda, L.K. , et al. 1997. Sensitization to *Blomia tropicalis* in patients with asthma and identification of allergen Blo t 5. *Am J Resp Crit Care Med* 155(1):342-350.

Asderakis, A., et al. 1997. CA repeat polymorphism in the first intron of the interferon-γ (IFN-γ) gene. *Immunol* 92(suppl.):62.

Barnes, K.C., et al. 1996. Linkage of asthma and total serum IgE concentration to markers on chromosome 12q: Evidence from Afro-Caribbean and Caucasian populations. *Genomics* 37:41-50.

Bender, W., P. Pierer, and D.S. Hogness. 1983. Chromosomal walking and jumping to isolate DNA from the Ace and rosy loci and the bithorax complex in *Drosophila melanogaster*. *J Mol Biol* 168:17-33.

Bernton, H.S., and H. Brown. 1970. Insect allergy: the allergenicity of the excrement of the cockroach. *Ann Allergy* 28:543-547.

Blumenthal, M., et al. 1992. HLA-DR2 [HLA-B7, SC31, DR2], and HLA-B8, SC01, DR3] haplotypes distinguish subjects with asthma from those with rhinitis only in ragweed pollen allergy. *J Immunol* 148(2):411-416.

Blumenthal, M.N., and D.B. Amos. 1987. Genetic and immunologic basis of atopic responses. *Chest* 91(June, 1987 supplement):176S-184S.

Boguniewicz, M. , and A. Hayward. 1996. Atopy, airway responsiveness, and genes. *Thorax* 51:S55-S59.

Bollinger, M, E. , et al. 1996. Cat antigen in homes with and without cats may induce allergic symptoms. *J Allergy Clinical Immunol* 97(4):907-914.

Borecki, I.B., D.C. Rao, J.M. Lalouel, and J.W. Gerrard. 1985. Demonstration of a common major gene with pleiotropic effects on immunoglobulin E levels and allergy. *Gen Epidemiol* 2:327.

Bradding, P., et al. 1993. Immunolocalisation of cytokines in the nasal mucosa of normal and perennial rhinitic subjects: the mast cell as a source of IL-4, IL-5 and IL-6 in human allergic inflammation. *J Immunol* 151:3853-3865.

Braman, S.S., et al. 1987. Airway hyperresponsiveness in allergic rhinitis: a risk factor for asthma. *Chest* 91:671-674.

Brenner, R.J., P.G. Koehler, and R.S. Patterson. 1987. Health implications of cockroach infestations. *Infections in Med* 4(8):349-359, 393.

Bruce, C.A., et al. 1975. Quantitative inhalation bronchial challenge in ragweed hay fever patients: a comparison with ragweed-allergic asthmatics. *J Allergy Clin Immunol* 56:331-337.

Burge, H.A. 1989. Airborne allergenic fungi. *Immunol Allergy Clins N Amer* 9:307-319.

Burrows, B., et al. 1989. Association of asthma with serum IgE levels and skin-test reactivity to allergens. *N Engl J Med* 320:271-277.

Burrows, B. , et al. 1995. Relations of bronchial responsiveness to allergy skin test reactivity, lung function, respiratory symptoms, and diagnoses in thirteen-year-old New Zealand children. *J Allergy Clin Immunol* 95(2):548-556.

Call, R.S. , et al.1992. Risk factors for asthma in inner city children. *J Peds* 121(6):862-866.

Campbell, D.A., et al. 1997. D12S366 is associated with asthma and atopy. *Am J Resp Crit Care Med 155(4)*:A701.

Caraballo, L. , et al. 1996. HLA-DR3 is associated with the IgE immune responsiveness to a recombinant allergen from *Blomia tropicalis* (BT). *Adv Exp Med Biol* 409:81-83.

Caraballo, L.R. , and M. Hernandez. 1990. HLA haplotype segregation in families with allergic asthma. *Tissue Antigens* 35(4):182-186.

Chan-Yeung, M. , et al. 1995. Mite and cat allergen levels in homes and severity of asthma. *Am J Resp Crit Care Med* 152(6 Pt 1):1805-1811.

Chapman, M.D. , et al. 1988. Monoclonal antibodies to the major feline allergen Fel d I. II. Single step affinity purification of Fel d I, N-terminal sequence analysis, and development of a sensitive two-site immunoassay to assess Fel d I exposure. *J Immunol* 140(3):812-818.

Chapman, M.D. , et al. 1987. Monoclonal immunoassays for major dust mite (Dermatophagoides) allergens, Der p I and Der f I, and quantitative analysis of the allergen content of mite and house dust extracts. *J Allergy Clin Immunol* 80(2):184-194.

Chapman, M.D. , et al.1997. Defined epitopes: in vivo and in vitro studies using recombinant allergens. *Intl Arch Allergy Immunol* 113(1-3):102-104.

Charpin, C. , et al. 1991. Fel d I allergen distribution in cat fur and skin. *J Allergy Clin Immunol* 88(1):77-82.

Collée, J.M., et al. 1993. Allele sharing on chromosome 11q13 in sibs with asthma and atopy. *Lancet* 342:936.

Colloff, M.J., et al. 1992. The control of allergens of dust mites and domestic pests: A position paper. *Clin Exp Allergy* 22:1-28.

Cookson, W., and J.M. Hopkin. 1988. Dominant inheritance of atopic immunoglobulin E responsiveness. *Lancet* 1:86.

Cookson, W., et al. 1989. Linkage between immunoglobulin E responses underlying asthma and rhinitis and chromosome 11q. *Lancet* 1:1292-1295.

Cookson, W.O.C.M., et al. 1992. Maternal inheritance of atopic responsiveness on chromosome 11q. *Lancet* 340:381-384.

Coulter, K.M., et al. 1987. Specific IgE antibody responses to ragweed allergens Ra5S and Ra5G associated with distinct HLA-DR genes. *Mol Immunol* 24:1207-1210.

Daniels, S.E., et al. 1996. A genome-wide search for quantitative trait loci underlying asthma. *Nature* 383:247-250.

Desjardins, A.C., et al. 1993. Exposure to domestic animals and risk of immunologic sensitization in subjects with asthma. *J Allergy Clin Immunol* 91(5):979-986.

Djukanovic, R., et al. 1990. Mucosal inflammation in asthma. *Am Rev Resp Dis* 142:434-457.

Doull, I.J.M., et al. 1996. Allelic association of gene markers on chromosomes 5q and 11q with atopy and bronchial hyperresponsiveness. *Am J Resp Crit Care Med* 153:1280-1284.

Einarsson, R., A.K.M. Munir, and S.K.G. Dreborg. 1995. Allergen in school dust: major mite (Der p 1, Der f 1) allergens in dust from Swedish schools. *J Allergy Clin Immunol* 95:1049-1053.

Ewens, W.J., and R.S. Spielman. 1995. The transmission/disequilibrium test: history, subdivision, and admixture. *Am J Hum Gen* 57:455-464.

Fernandez-Caldas, E. , et al. 1993a. Cutaneous sensitivity to six mite species in asthmatic patients from five Latin American countries. *J Invest Allergol Clin Immunol* 3(5):245-249.

Fernandez-Caldas, E. , et al. 1993b. Mite fauna, Der p I, Der f I and Blomia tropicalis allergen levels in a tropical environment. *Clin Exp Allergy* 23(4):292-297.

Ferrandiz, R., R. Casas, and S. Dreborg. 1996. Sensitization to *Dermatophagoides siboney, Blomia tropicalis*, and other domestic mites in asthmatic patients. *Allergy* 51(7):501-505.

Ford, A.W. , L. Alterman, and D.M. Kemeny. 1989. The allergens of dog. I. Identification using crossed radio-immunoelectrophoresis. *Clin Exp Allergy* 19(2):183-90, 1989 Mar. 19(2):183-190.

Ford, A.W. , and D.M. Kemeny. 1992. The allergens of dog. II. Identification and partial purification of a major dander allergen. *Clin Exp Allergy* 22(8):793-803.

Freidhoff, L.R., et al. 1988. Association of HLA-DR3/Dw3 and total serum immunoglobulin E level with human immune response to *Lol p* I and *Lol p* II allergens in allergic subjects. *Tissue Antigens* 31:211.

Freidhoff, L.R., and D.G. Marsh. 1993. Relationship among asthma, serum IgE levels and skin test sensitivity to inhaled allergens. *Intl Arch Allergy Immunol* 100:355-361.

Gauchat, J-F., et al. 1993. Induction of human IgE synthesis in B cells by mast cells and basophils. *Nature* 365:340-343.

Gelber, L.E., et al. 1993. Sensitization and exposure to indoor allergens as risk factors for asthma among patients presenting to hospital. *Am Rev Resp Dis* 147:573-578.

Gerrard, J.W., D.C. Rao, and N.E. Morton. 1978. A genetic study of immunoglobulin E. *Am Journal Hum Gen* 30:46-58.

Hamilton, R.G., et al. 1992. House dust aeroallergen measurements in clinical practice: a guide to allergen-free home and work environments. *Immunol Allergy Pract* 14:96-112.

Haseman, J.K., and R.C. Elston. 1972. The investigation of linkage between a quantitative trait and a marker locus. *Behavioral Genetics* 2:3-19.

Helm, R.M., et al. 1994. Isolation of the 36-kD German (*Blattella germanica*) cockroach allergen using fast protein liquid chromatography. *Intl Arch Allergy Immunol* 103(1):59-66.

Heymann, P.W. , et al. 1989. Antigenic and structural analysis of group II allergens (Der f II and Der p II) from house dust mites (*Dermatophagoides spp*). *J Allergy Clin Immunol* 83(6):1055-1067.

Hill, M.R. , and W.O. Cookson. 1996. A new variant of the beta subunit of the high-affinity receptor for immunoglobulin E (Fc epsilon RI-beta E237G): associations with measures of atopy and bronchial hyperresponsiveness. *Hum Mol Gen* 5(7):959-962.

Hizawa, N., et al. 1998a. Linkage analysis of *Dermatophagoides pteronyssinus* (Der p)-specific IgE responsiveness with polymorphic markers on chromosome 6p21 (HLA-D region) in Caucasian families by Transmission/Disequilibrium Test (TDT). *J Allergy Clin Immunol* 102:443-448.

Hizawa, N., et al. 1998b. Genetic regulation of *Dermatophagoides pteronyssinus* (Der p)-specific IgE responsiveness: A genome-wide multipoint linkage analysis in families recruited through two asthmatic sibs. *J Allergy Clin Immunol* 102:436-442.

Hizawa, N., et al. 1998c. Genetic influences of chromosomes 5q31-q33 and 11q13 on specific IgE responsiveness to common inhaled allergens among African American families. *J Allergy Clin Immunol* 102:449-453.

Hizawa, N. , et al. 1995. Association between high serum total IgE levels and D11S97 on chromosome 11q13 in Japanese subjects. *J Med Gen* 32(5):363-369.

Horowitz, E., et al. 1996. Allergy to the *Blattella germanica* cockroach: a major risk factor for asthma severity in adolescents. *J Allergy Clin Immunol* 97:379.

Huang, S.K., P. Zwollo, and D.G. Marsh. 1991. Class II MHC restriction of human T-cell responses to short ragweed allergen, *Amb a* V. *Eur J Immunol* 21:1469-1473.

Ingram, J.M., et al. 1995. Quantitative assessment of exposure to dog (Can f 1) and cat (Fel d 1) allergens: Relation to sensitization and asthma among children living in Los Alamos, New Mexico. *J Allergy Clin Immunol* 96:449-456.

Ismaïl, A. , et al. 1997. Polymorphism in transporter antigen peptides gene (TAP1) associated with atopy in Tunisians. *J Allergy Clin Immunol*. 99(2):216-23 99(2):216-223.

Kang, B. 1976. Study on cockroach antigen as a probable causative agent in bronchial asthma. *J Allergy Clin Immunol* 58:357-365.

Kang, B. , and J.L. Chang. 1985. Allergenic impact of inhaled arthropod material. *Clin Rev Allergy* 3:363-375.

Kang, B.C. , J. Johnson, and C. Veres-Thorner. 1993. Atopic profile of inner-city asthma with a comparative analysis on the cockroach-sensitive and ragweed-sensitive subgroups. *J Allergy Clin Immunol* 92(6):802-811.

Kang, B.C., C.W. Wu, and J. Johnson. 1992. Characteristics and diagnoses of cockroach-sensitive bronchial asthma. *Ann Allergy* 68:237-244.

Kapsali, T., et al. 1997. Rhinitis is ubiquitous in allergic asthmatics. *J Allergy Clin Immunol* 99:S138.

Kivity, S. , et al. 1993. Mite asthma in childhood: a study of the relationship between exposure to house dust mites and disease activity. *J Allergy Clin Immunol* 91(4):844-849.

Konieczny, A., et al. 1997. The major dog allergens, Can f 1 and Can f 2, are salivary lipocalin proteins: cloning and immunological characterization of the recombinant forms. *Immunol* 92(4):577-586.

Krauter, K., et al. 1995. A second-generation YAC contig map of human chromosome 12. *Nature* 377(Supp):321-333.

Lander, E.S., and N.J. Schork. 1994. Genetic dissection of complex traits. *Science* 265:2037-2048.

Lehrer, S.B. , et al. 1994. Prevalence of basidiomycete allergy in the USA and Europe and its relationship to allergic respiratory symptoms. *Allergy*. 49(6) 49(6):460-465.

Leitermann, K. , and J.L. Ohman, Jr. 1984. Cat allergen 1: Biochemical, antigenic, and allergenic properties. *J Allergy Clin Immunol* 74(2):147-153.

Loveridge, J.A. , et al. 1991. The genetic contribution to human T-cell receptor repertoire. *Immunol* 74(2):246-250.

Lympany, P., et al. 1992. Genetic analysis of the linkage between chromosome 11q and atopy. *Clin Exp Allergy* 22:1085-1092.

Marks, G.B., et al. 1995. Mite allergen (Der p 1) concentration in houses and its relation to the presence and severity of asthma in a population of Sydney schoolchildren. *J Allergy Clin Immunol* 96(4):441-448.

Marsh, D.G. 1994. Genetics of atopy and IgE. In Samter's *Immunologic Diseases*. Pp. 1257-1272, Vol. II: Little, Brown & Co.

Marsh, D.G., W.B. Bias, and K. Ishizaka. 1974. Genetic control of basal serum immunoglobulin E level and its effect on specific reaginic sensitivity. *Procs Nat Acad Sci* 71:3588-3592.

Marsh, D.G., and M.N. Blumenthal. 1990. *Genetic and Environmental Factors in Clinical Allergy*. Minneapolis: University of Minnesota Press.

Marsh, D.G., A. Lockhart, and S.J. Holgate. 1993. *The Genetics of Asthma*. Oxford: Blackwell Scientific Publications.

Marsh, D.G., D.A. Meyers, and W.B. Bias. 1981. The epidemiology and genetics of atopic allergy. *N Engl J Med* 305(26): 1551-1559.

Marsh, D.G., et al. 1994. Linkage analysis of *IL4* and other chromosome 5q31.1 markers and total serum immunoglobulin E concentrations. *Science* 264:1152-1156.

Marsh, D.G., et al. 1990. Molecular studies of human response to allergens. *Cold Spring Harbor Symp on Quant Biol* 54:459-470.

Mata, P., et al. 1992. Fel d I allergen: skin and or saliva? *Ann Allergy* 69(4):321-322.

Meyers, D.A., et al. 1994. Evidence for a locus regulating total serum IgE levels mapping to chromosome 5. *Genomics* 23:464-470.

Moffatt, M.F., and W.O. Cookson. 1997a. Linkage and candidate gene studies in asthma. *Am J Resp Crit Care Med* 156(4 Pt 2):S110-112.

Moffatt, M.F., and W.O. Cookson. 1997b. Tumour necrosis factor haplotypes and asthma. *Hum Mol Gen* 6(4):551-554.

Moffatt, M.F., et al. 1994. Genetic linkage of T-cell receptor a/d complex to specific IgE responses. *Lancet* 343:1597-1600.

Moffatt, M.F. , et al. 1997. Germline TCR-A restriction of immunoglobulin E responses to allergen. *Immunogenetics* 46(3):226-230.

Mollet, J.A. , et al. 1997. Evaluation of German cockroach (*Orthoptera:Blattellidae*) allergen and seasonal variation in low-income housing. *J Med Entomol* 34(3):307-311.

Mossman, T.R., and R.L. Coffman. 1989. Heterogeneity of cytokine secretion patterns and functions of helper T cells. *Adv Immunol* 46:111-147.

Munir, A.K.M., et al. 1993. Allergens in school dust: the amount of the major cat (Fel d 1) and dog (Can f 1) allergens in dust from Swedish schools is high enough to probably cause perennial symptoms in most children with asthma who are sensitized to cat and dog. *J Allergy Clin Immunol* 91:1067-1074.

Munir, A.K.M., N-I.M. Kjellman, and B. Bjorksten. 1997. Exposure to indoor allergens in early infancy and sensitization. *J Allergy Clin Immunol* 100:177-181.

Neffen, H.E. , et al. 1996. Mite sensitivity and exposure in the city of Santa Fe, Argentina. *J Invest Allergol Clin Immunol* 6(5):278-282.

Nelson, R.P. Jr. , et al. 1996. Allergen-specific IgE levels and mite allergen exposure in children with acute asthma first seen in an emergency department and in nonasthmatic control subjects. *J Allergy Clin Immunol* 98(2):258-263.

Nickel, R., et al. 1997. Evidence for linkage of chromosome 12q15-q24.1 markers to high total serum IgE concentrations in children of the German Multicenter Allergy Study (MAS '90). *Genomics* 46:159-162.

O'Hehir, R.E., et al. 1991. The specificity and regulation of T-cell responsiveness to allergens. *Ann Rev Immunol* 9:67-95.

Ohman, J.L. Jr. , and F.C. Lowell. 1977. IgE antibody to cat allergens in an allergic population. *J Allergy Clin Immunol* 60(5):317-323.

Okano, M. , et al. 1996. Epitope analysis of HLA-DR-restricted helper T-cell responses to Der p II, a major allergen molecule of *Dermatophagoides pteronyssinus*. *Allergy* 51(1):29-35.

Paul, W.E., and R.A. Seder. 1994. Lymphocyte responses and cytokines. *Cell* 76:241.

Peat, J.K. , et al. 1994. Asthma severity and morbidity in a population sample of Sydney schoolchildren: Part II--Importance of house dust mite allergens. *Aust & NZ J Med* 24(3):270-276.

Peat, J.K., et al. 1996. House dust mite allergens: a major risk factor for childhood asthma in Australia. *Am J Resp Crit Care Med* 153:141-146.

Platts-Mills, T.A.E. , et al. 1995a. Dog and cat allergens and asthma among school children in Los Alamos, New Mexico, USA: altitude 7,200 feet. *Intl Arch Allergy Immunol* 107(1-3):301-303.

Platts-Mills, T.A.E., et al. 1997. Indoor allergens and asthma: report of the Third International Workshop. *J Allergy Clin Immunol* 100(6 Pt 1):S2-24.

Platts-Mills, T.A.E., and M.C. Carter. 1997. Asthma and indoor exposure to allergens. *N Engl J Med* 336:1382-1384.

Platts-Mills, T.A.E., et al. 1985. Mites of the genus *Dermatophagoides* in dust from the houses of asthmatic and other allergic patients in North America: Development of a radioimmunoassay for allergen produced by *D. farinae* and *D. pteronyssinus*. *Intl Arch Appl Immunol* 77:163.

Platts-Mills, T.A.E., et al. 1992. Dust mite allergens and asthma: Report of a second international workshop. *J Allergy Clin Immunol* 89:1046-1060.

Platts-Mills, T.A.E., et al. 1995b. Is there a dose-response relationship between exposure to indoor allergens and symptoms of asthma? *J Allergy Clin Immunol* 96:435-440.

Platts-Mills, T.A.E., and A.L. de Weck. 1989. Dust mite allergens and asthma--A worldwide problem. *J Allergy Clin Immunol* 83:416-427.

Pollart, S.M., et al. 1989. Epidemiology of acute asthma: IgE antibodies to common inhalant allergens as a risk factor for emergency room visits. *J Allergy Clin Immunol* 83(5):875-882.

Pollart, S.M. , et al. 1991a. Identification, quantitation, and purification of cockroach allergens using monoclonal antibodies. *J Allergy Clin Immunol* 87(2):511-521.

Pollart, S.M., et al. 1991b. Environmental exposure to cockroach allergens:

Analysis with monoclonal antibody-based enzyme immunoassays. *J Allergy Clin Immunol* 87:505-510.

Postma, D.S., et al. 1995. Genetic susceptibility to asthma --bronchial hyperresponsiveness coinherited with a major gene for atopy. *N Engl J Med* 333:894-900.

Puerta, L. , et al. 1996. Sequential determinations of *Blomia tropicalis* allergens in mattress and floor dust samples in a tropical city. *J Allergy Clin Immunol* 97(2):689-691.

Risch, N. 1990. Linkage strategies for genetically complex traits. Parts I, II and III. *Amer J Hum Gen* 46:229-241.

Roebber, M., et al. 1985. Immunochemical and genetic studies of *Amb. t.* V (Ra5G), an Ra5 homologue from giant ragweed pollen. *J Immunol* 134:3062.

Romagnani, S. 1992. Human TH1 and TH2 subsets: Regulation of differentiation and role in protection and immunopathology. *Intl Arch Allergy Immunol* 98:279-285.

Rosenstreich, D.L., et al. 1997. The role of cockroach allergy and exposure to cockroach allergen in causing morbidity among inner-city children with asthma. *N Eng J Med* 336:1356-1363.

Rosenwasser, L.J., et al. 1995. Promoter polymorphisms in the chromosome 5 gene cluster in asthma and atopy. *Clin Exp Allergy* 25(Supplement 2):74-78.

Saltman, D.L., et al. 1993. A physical map of 15 loci on human chromosome 5q23–q33 by two-color fluorescence *in situ* hybridization. *Genomics* 16:726-732.

Sarpong, S.B., et al. 1997. Cockroach allergen (Bla g 1) in school dust. *J Allergy Clin Immunol* 99(4):486-492.

Schou, C., et al. 1991a. Assay for the major dog allergen, Can f I: investigation of house dust samples and commercial dog extracts. *J Allergy Clin Immunol* 88(6):847-853.

Schou, C., et al. 1990. Identification and purification of an important cross-reactive allergen from American (*Periplaneta americana*) and German (*Blattella germanica*) cockroach. *J Allergy Clin Immunol* 86(6 Pt 1):935-946.

Schou, C. , U.G. Svendsen, and H. Lowenstein. 1991b. Purification and characterization of the major dog allergen, Can f I. *Clin Exp Allergy* 21(3):321-328.

Sears, M.R., et al. 1991. Relation between airway responsiveness and serum IgE in children with asthma and in apparently normal children. *N Engl J Med* 325:1067-1071.

Shirakawa, T., et al. 1994a. Linkage between severe atopy and chromosome 11q13 in Japanese families. *Clin Gen* 46:228-232.

Shirakawa, T., et al. 1994b. Association between atopy and variants of the β subunit of the high-affinity immunoglobulin E receptor. *Nature*

Genetics 7:125.

Smith, J.M. 1983. Epidemiology and natural history of asthma, allergic rhinitis, and atopic dermatitis (eczema). In *Allergy: Principles and Practice*. E.M. Jr., C.E. Reed, and E.F. Ellis, eds. Pp. 771-803. St. Louis: Mosby.

Spies, T. , et al. 1990. A gene in the human major histocompatibility complex class II region controlling the class I antigen presentation pathway. *Nature* 348(6303):744-747.

Sporik, R., et al. 1990. Exposure to house-dust mite allergen (*Der p* I) and the development of asthma in childhood. *N Engl J Med* 323:502-507.

Sporik, R., et al. 1995. Association of asthma with serum IgE and skin test reactivity to allergens among children living at high altitude: tickling the dragon's breath. *Am J Resp Crit Care Med* 151:1388-1392.

Squillace, S.P. , et al. 1997. Sensitization to dust mites as a dominant risk factor for asthma among adolescents living in central Virginia. Multiple regression analysis of a population-based study. *Am J Resp Crit Care Med* 156(6):1760-1764.

Tollerud, D.J., et al. 1991. Asthma, hay fever and phlegm production associated with distinct patterns of allergy skin test reactivity, eosinophilia, and serum IgE levels. *Am Rev Resp Dis* 144:775-781.

Troutman, C.M. 1988a. Cat owners and their use of veterinary services. *J Amer Vet Med Assoc* 193(10):1217-1219.

Troutman, C.M. 1988b. Dog owners and their use of veterinary services. *J Amer Vet Med Assoc* 193(9):1056-1058.

Tsai, J.J. , et al. 1998. Sensitization to *Blomia tropicalis* among asthmatic patients in Taiwan. *Intl Archives Allergy Immunol* 115(2):144-149.

Vercelli, D., and R.S. Geha. 1991. Regulation of IgE synthesis in humans: a tale of two signals. *J Allergy Clin Immunol* 88:285-295.

Vervloet, D. , et al. 1991. Medication requirements and house dust mite exposure in mite-sensitive asthmatics. *Allergy* 46(7):554-558.

Voorhorst, R., F.Th.M. Spieksma, and N. Varekamp. 1969. *House dust mite atopy and the house dust mite Dermatophagoides pteronyssinus* (Troussart, 1897). Leiden: Stafleu's Scientific Publishing Company.

Wahn, U., et al. 1997. Indoor allergen exposure is a risk factor for sensitization during the first three years of life. *J Allergy Clin Immunol* 99:763-769.

Walley, A.J., et al. 1997. Linkage and allelic association of chromosome 5 microsatellite markers with atopic asthma phenotypes in a general population sample. *Amer J Resp Crit Care Med* 155(April):A257.

Wedderburn, L.R. , et al. 1993. *In vivo* clonal dominance and limited T-cell receptor usage in human CD4+ T-cell recognition of house dust mite allergens. *Procs Natl Acad Sci USA* 90(17):8214-8218.

Wickman, M. , et al. 1991. House dust mite sensitization in children and residential characteristics in a temperate region. *J Allergy Clin Immunol* 88(1):89-95.

Wilkinson, J., et al. 1997. Evidence for linkage between markers on chromosome 12q and asthma and atopy in 191 families. *Amer J Resp Crit Care Med* 155(4):A257.

Wilkinson, J., et al. 1996. Evidence for linkage for atopy and asthma to markers on chromosome 12q. *Eur Resp J* 9:435S.

Wood, R.A. , et al. 1989. The effect of cat removal on allergen content in household-dust samples. *J Allergy Clin Immunol* 83(4):730-734.

Woods, F.E., et al. 1981. Survey of Attitudes and Knowledge of Public Housing Residents Toward Cockroaches. *Bull Entomol Soc Amer* 27(1):9-13.

Young, R.P., et al. 1992. Confirmation of genetic linkage between atopic IgE responses and chromosome 11q13. *J Med Gen* 29(4):236-238.

Zwollo, P., E. Ehrlich-Kautzky, A.A. Ansari, S.J. Scharf, H.A. Erlich, and D.G. Marsh. 1991. Molecular studies of human immune response genes for the short ragweed allergen, Amb a V. Sequencing of HLA-D second exons in responders and non-responders. *Immunogenetics* 33:141-151.

Chapter 11

THE CONTRIBUTION OF GENES AND ENVIRONMENT TO ASTHMA

Lanny J. Rosenwasser, M.D.

I. INTRODUCTION

The prevalence, morbidity and mortality of asthma has increased worldwide over the past three decades. This disturbing increase in this now common condition has occurred in spite of much greater understanding of the underlying immunopathogenesis of bronchial asthma and the introduction of new and safer treatments for the disease. It is clear that the epidemiologically confirmed increases in prevalence and mortality are multifactorial. Asthma is a condition that not only reflects underlying host predisposition to development of the disease but also the response of the host to environmental stimuli. Thus, we need to consider the relative contributions of both of these factors in the generation of this disease. That the environmental assault on the asthmatic phenotype has increased in the past three decades is likely but should not be accepted without some degree of modification. The increasing prevalence and awareness concerning the deteriorating quality of indoor air, especially from the point of view of contaminating allergens, suggests that perhaps these factors have played an overwhelming role in the generation of increased risk. Outdoor air quality, another factor that has been considered on the etiology of bronchial asthma, also may be playing a role in the generation of this problem. On the other hand the prevalence of asthma has increased in areas such as New Zealand where outdoor air quality has always been good or in London where air quality has improved. The possibility that genetic contributions have changed in the past three decades is not likely. However, as new populations have been exposed to more and more assaults in terms of indoor and outdoor air quality, this may have altered the overall prevalence of the disease as differing susceptibilities (e.g. different ethnicity) have become involved. The present short review will examine a number of environmental factors that may be related to asthma and will also examine issues related to host defense. The contribution of genes and the environment to asthma is not a zero sum game in that it is quite possible that as the environmental assault has increased over the past few decades, thresholds for recruitment of genes may have been significantly altered by the asthmatic phenotype. We are studying a complex multigenic disease, the expression of which is influenced by multiple environmental factors, not only allergens but also chemical air pollution, viruses and tobacco smoke.

II. ENVIRONMENT

Clearly a number of important factors contribute to environmental issues

1-56670-373-5/99/$0.00+$.50

in those individuals who are genetically susceptible to developing asthma and allergic sensitization. One of the best paradigms for studying the potential contribution of environmental responses to allergic disease has come from studies that compare industrialized with less industrialized areas and rural with urban areas. Such studies clearly demonstrate that Westernized, more industrial societies over the past 20-30 years have a higher rate of atopic sensitization and higher prevalence of asthma, compared to those less modern countries characterized by many of the countries of the former Eastern Bloc (von Mutius, 1994). In addition, urban (as opposed to rural) dwelling seems to have a greater risk of allergen-related sensitization in terms of atopy and asthma, suggesting that outdoor pollution, while not an overriding factor, may nonetheless be an important contributor. These facts have led to the recognition that there may be four or more potential contributors to the circumstances that lead to higher rates of sensitization. As we will discuss in more detail, indoor air quality is probably the overriding factor but there are other factors that are involved including outdoor air quality and air pollution where exposure to particulate matter and diesel exhaust and others may enhance immune reactivity and allergen reactivity both in the nose and in the lower airway (Nel *et al,* 1998; Pope, 1994; Lang *et al*, 1994). Exposure to infection may also play a role both from the point of view of potential protectiveness against atopy via shift of immune responses to a Th1 paradigm or by appropriate sensitization of the immune system in a specific developmental time period. Data looking at BCG sensitization as a factor protecting against the development of asthma and other circumstances that could in some way influence Th1 immune cell reactivity, have also been reported (Shirakawa *et al*, 1997). In this same vein, different prevalence of viral infections may explain epidemiologic findings that suggest that sib-ship and place within the family is associated with greater or lesser degrees of risk for sensitivity and allergic disease. Results of this kind have come from family studies in the UK, as well as large recruit studies done in Italy (Butland *et al*, 1997). The influence of potential dietary changes on these responses should also be considered as gut flora and sensitization seem to correlate (Bjorksten *et al*, 1997). These dietary influences may be profoundly influenced by antibiotic usage and other circumstances that can lead to significant changes in exposures and changes in the handling or the allergens that are involved in atopic sensitization. Finally, a sedentary lifestyle with lack of physical activity has also been implicated as a contributor to the increasing rates of asthma.

The role of indoor allergens is especially intriguing. The major indoor allergens include house dust mites in those locations where mite growth is enhanced, animal dander derived from indoor warm-blooded pets, molds which can also occur indoors and cockroach allergen which has recently been recognized as an important contributor to asthma sensitivity. Exposure to indoor allergens is very critical to this issue as some have suggested that levels of certain antigens in certain areas may correlate with the risk for both sensitization and risks of exacerbation of the disease (Third International Workshop, 1997). Another issue relevant to the worsening indoor air quality has been the fact that tighter houses built with fewer air exchanges per day and decreased ventilation as a means of economically approaching the energy crisis of the past two to three decades, has led to greater increases in ambient levels of allergens. Keeping homes heated and at a comfortable humidity also encourages mite growth as does the use of fitted carpets, the increased use of wall coverings, draperies and furniture that are potential reservoirs for animal dander and other indoor allergens including mite and cockroach (Platts-Mills & Chapman, 1987). All of these indoor allergen problems may have contributed to the environmental assault and the increasing prevalence of asthma.

III. THE ROLE OF GENETICS IN ASTHMA

Asthma has a significant genetic component without any question. There are circumstances that lead to enhanced immune reactivity, to those allergens that are inhaled, that seem to correlate with the production of IgE and the development of airway hyperresponsiveness. The intensity of these immune reactions clearly has been correlated with the intensity of hyperresponsiveness in asthma. Furthermore, the genetic contribution to this profile has been well documented over the past two decades, beginning with twin studies that identified these problems and attributed about a 50-50 split to the contributions of the environment and genetics to asthma. While the gene pool has not changed significantly in the past two decades, genes that are recruited especially in a polygenic disease such as asthma are probably associated with the increased prevalence of disease. It is not so much that more genes are now susceptible for asthma, but rather that the threshold for the polygenic, multigenic susceptibility may have been altered by environmental assaults in the past two and one-half decades on atopic sensitization and the associated tendency to asthma. In fact, whole genome screens and studies of single nucleotide polymorphisms for potential candidate genes all identify this multigenic complex disease approach that correlates with asthma susceptibility. Multiple genes within the same locus, identified using candidate and positional cloning strategies as potential loci for linkages and associations, have identified a number of important genes that may contribute to the overall asthmatic syndrome. These include genes that control for allergen and antigen reactivity in IgE production and T cell activation and eosinophil, mast cell, and basophil function and a set of genes that are involved in generally maintaining inflammation and potential bronchial hyperreactivity. In a complex syndrome such as atopy or asthma, the predominance of a single gene or even the overriding influence of a few genes is unlikely. A more likely scenario involves the interaction of multiple sets of genes, or polygenes, each contributing a small amount to the pathophysiology.

To date, three types of studies of the genetics of asthma and atopy have been done. The first involves the use of candidate genes for linkage or association studies in families or populations of patients to establish whether the genes are involved in the disease (Cookson *et al*, 1989; Doull *et al*, 1996; Meyers *et al*, 1994; Moffatt *et al*, 1994; Postma *et al*, 1995; Young *et al*, 1994). Typical examples of candidate genes are the IL-4 and IL-5 promoter genes and the gene for the high affinity IgE receptor. A second approach uses DNA microsatellite markers to scan the entire genome for linkages associated with asthma in families. Two such studies have been reported in the past year and have identified multiple linkages within the genome, each of which may be associated with attractive candidate genes (Daniels *et al*, 1996; The Collaborative Study, 1997). The overall goal of this type of positional cloning is to identify previously undiscovered genes that might be associated with disease. These studies have uncovered at least 14 or 15 separate linkages, some of which may turn out to be false positive results. Nonetheless, multiple genes are probably involved in the asthmatic syndrome, each of which may make a small contribution to the total genetic influence of the disease. It is assumed that each of these genes acting together with environmental exposure can make important contributions to the development of the syndrome.

The third type of study of the genetics of asthma identifies structural and functional associations with polymorphisms or mutations in the DNA or protein structure of candidate genes or their promoter regions. In addition to the interleukin-4 gene studies (Borish *et al*, 1994; Deichmann *et al*, 1998; Hershey *et al*, 1997; Marsh *et al*, 1994; Pritchard *et al*, 1991; Rosenwasser *et al*, 1995), examples include studies that have identified alternative forms of the β chain of the high-affinity receptor (FcϵRIβ) and structural variants in the β-adrenergic-receptor gene (Hall *et al*, 1995; Shirakawa *et al*, 1994; Turki *et al*, 1995).

The genetic studies are summarized in Table 1. Each of these genes appears to contribute to the susceptibility to asthma although usually not more than a couple of percent. Therefore, the likelihood of a genetic approach to altering susceptibility to asthma via gene therapy is remote. A more likely scenario would involve the development of techniques for pharmacogenetic manipulation, and the characterization of individuals who are more or less likely to respond to a certain medicine. Hence, asthmatics may [on the basis of their genotype, for the 5-lipoxygenase genes] have an enhanced bronchodilator response to a lipoxygenase inhibitor. Individuals who have certain genotypes identifying high output IL-4 or high IL-4 receptor activity may be more those in whom inhibition of cytokines such as IL-4 and IL5 could be a therapeutic target for treatment. The genotype of the beta-adrenergic receptor may dictate the best type of bronchodilatation that can be applied to an individual who has a certain genetic susceptibility profile at the beta-adrenergic receptor locus. All of these approaches will lead to a more focused and a better treatment profile for asthmatics based on their particular genetic susceptibility. This hypothesis will need to be tested in the next years.

Table 1: Candidate Genes in Atopy/Asthma

Class/Function	Gene	Chromosomol Location
Recognition	MHC TCRVα	6p 7p
Signaling	STAT6	12q
Cytokines	IL-3, IL-4, IL-5, IL-9 GM-CSF, IL-12β	5q
Cytokine Receptors and Other Receptors	IL-4Rα Glucocorticoid receptor β2 Adrenergic Receptor	16p 5q 5q
Transcription Factors	NFAT-1 NFAT-4 C-maf Egr	20q 16q 16q 5q
Inflammation Genes	iNOS LTC4 synthase	12q 5q

References

Borish, L., Mascali, J.J., Klinnert, M., Leppert, M., Rosenwasser, L.J. SSC polymorphisms in interleukin genes. *Hum. Mol. Genet.*, 3, 1710, 1994. [Erratum, *Hum. Mol. Genet.* 4, 974, 1995]

Butland, B.K., Strachan, D.P., Lewis, S., Brynner, J., Butler, N., Britton, J. Investigation into the increase in hay fever and eczema at age 16 observed between the 1958 and 1970 British birth cohorts. *Brit. Med. J.*, 315, 717-721, 1997.

Cookson, W.O., Sharp, P.A., Faux, J.A., Hopkin, J.M. Linkage between immunoglobulin E responses underlying asthma and rhinitis and chromosome 11q. *Lancet*, 1, 1291-1295, 1989.

Daniels, S.E., Bhattacharrya, S., James, A., et al. A genome-wide search for quantitative trait loci underlying asthma. *Nature*, 383, 247-250, 1996.

Deichmann, K.A., Heinzmann, A., Forster, J., et al. Linkage and allelic association of atopy and markers flanking the IL-4 receptor gene. *Clin. Exp. Allergy*, 18, 151-155, 1998.

Doull, I.J., Lawrence, S., Watson, M., et al. Allelic association of gene markers on chromosomes 5q and 11q with atopy and bronchial hyperresponsiveness. *Am. J. Respir. Crit. Care Med.*, 153, 1280-1284, 1996.

Hall, I.P., Wheatley, A., Wilding, P., Liggett, S.B. Association of Glu 27 β_2-adrenoceptor polymorphism with lower airway reactivity in asthmatic subjects. *Lancet,* 345, 1213-1214, 1995.

Hershey, G.K.K., Friedrich, M.F., Esswein, L.A., Thomas, M.L., Chatila, T.A. The association of atopy with a gain-of-function mutation in the α subunit of the interleukin-4 receptor. *N. Engl. J. Med.*, 337, 1720-1725, 1997.

Lang, D.M., Polansky, M. Patterns of asthma mortality in Philadelphia from 1969 to 1971. *N. Engl. J. Med.*, 331, 1542-1546, 1994.

Marsh, D.G., Neely, J.D., Breazeale, D.R., et al. Linkage analysis of IL-4 and other chromosome 5q31.1 markers and total serum immunoglobulin E concentrations. *Science*, 264, 1152-1156, 1994.

Meyers, D.A., Postma, D.S., Panhuysen, C.I., et al. Evidence for a locus regulating total serum IgE levels mapping to chromosome 5. *Genomics*, 23, 464-470, 1994.

Moffatt, M.F., Hill, M.R., Cornelis F., et al. Genetic linkage of T-cell receptor α/γ complex to specific IgE responses. *Lancet*, 343, 1597-1600, 1994.

Nel, A.E., Diaz-Sanchez, Ng, D., Hiura, T., Saxon, A. Enhancement of allergic inflammation by the interaction between diesel exhaust particles and the immune system [Editorial]. *J. Allergy Clin. Immunol.*, 102, 539-557, 1998.

Platts-Mills, T.A.E., Chapman, M.D. Dust mites: Immunology, allergic disease, and environmental control [Review]. *J. Allergy Clin. Immunol.,* 80, 755-775, 1987.

Pope, C.A. Respiratory hospital admissions associated with PM10 pollution in Utah, Salt Lake and Cache Valleys. *Arch. Envir. Health*, 46, 90-97, 1991.

Postma, D.S., Bleeker, E.R., Amelung, P.J., et al. Genetic susceptibility to asthma -- bronchial hyperresponsiveness coinherited with a major gene for atopy. *N. Engl. J. Med.*, 333, 894-900, 1995.

Pritchard, M.A., Baker, E., Whitmore, S.A., et al. The interleukin-4 receptor gene (*IL4R*) maps to 16p11.2-16p12.1 in human and to the distal region of mouse chromosome 7. *Genomics*, 10, 801-806, 1991.

Rosenwasser, L.J., Klemm, D.J., Dresback, J.K., et al. Promoter polymorphisms in the chromosome 5 gene cluster in asthma and atopy. *Clin. Exp. Allergy*, 25, 74-78, 1995.

Shirakawa, T.S., Li, A., Dubowitz, M., et al. Association between atopy and variants of the β subunit of the high-affinity immunoglobulin E receptor. *Nat. Genet.*, 7, 125-129, 1994.

Shirakawa, T., Enomoto, T., Shimazu, S., and Hopkin, J.M. The inverse association between tuberculin responses and atopic disorder. *Science*, 275, 77-9, 1997.

The Collaborative Study on the Genetics of Asthma. A genome-wide search for asthma susceptibility loci in ethnically diverse population. *Nat. Genet.*, 15, 389-392, 1997.

Third International Workshop, Cuenca, Spain. Co-Chairmen: Platts-Mills, T.A.E., Vervloet, D., Thomas, W.R., Aalberse, R.C., Chapman, M.D. Indoor allergens and asthma. *J. Allergy Clin. Immunol.*, 100, S1-S24, 1997.

Turki, J., Pak, J., Green, S.A., Martin, R.J., Liggett, S.B. Genetic polymorphisms of the β2-adrenergic receptor in nocturnal and non-nocturnal asthma: Evidence that Gly 16 correlates with the nocturnal phenotype. *J. Clin. Invest.*, 95, 1635-1641, 1995.

Von Mutius, E., Martinez, F.D., Fritzsch, C., Nicolai, R., Roell, C., Thiemann, H.H. Prevalence of asthma and atopy in two areas of West and East Germany. *Am. J. Respir. Crit. Care Med.*, 149, 358-64, 1994.

Young, R.P., Dekker, J.W., Wordsworth, B.P., et al. HLA-DR and HLA-DP genotypes and immunoglobulin E responses to common major allergens. *Clin. Exp. Allergy*, 24, 431-439, 1994.

Index

B

C

D

E

F

G

H

I

L

M

N

O

P

R

S

T

U

V

W